When the Rivers Run Dry

When the Rivers Run Dry

Water—The Defining Crisis of the Twenty-first Century

FRED PEARCE

Beacon Press
Boston

Beacon Press
25 Beacon Street
Boston, Massachusetts 02108-2892
www.beacon.org

Beacon Press books
are published under the auspices of
the Unitarian Universalist Association of Congregations.

10 09   8 7 6

This book is printed on acid-free paper that meets the uncoated paper
ANSI/NISO specifications for permanence as revised in 1992.

Maps by Hardlines Studio, UK
Text design by Patricia Duque Campos
Composition by Wilsted & Taylor Publishing Services

Library of Congress Cataloging-in-Publication Data

Pearce, Fred.
When the rivers run dry: water—the defining crisis of the twenty-first century /
Fred Pearce.
    p. cm.
Includes index.
ISBN-13: 978-0-8070-8573-8 (pbk. : alk. paper)   1. Water resources development.
2. Water-supply.   I. Title.

TC405.P43 2006
333.91—dc22   2005027495

For Joe, another river
who died before his time

Contents

Introduction

It would be wrong to say that I have been obsessed with rivers all my life. But, much as when taking a drive down a long river valley, I can't seem to avoid coming back to them. They have a hold on me.

I was brought up in a village in southeastern England where two small rivers begin their journey to the sea. One, the River Len, flowed west to the Medway and the Thames estuary. It powered a mill, I remember, and filled a lake in a park. Then one day it flooded and I couldn't get to school. That struck me: the power of a river unleashed.

The other river, the Stour, flowed east through the cathedral city of Canterbury, took in some extra water bubbling up from the bowels of the East Kent coal mines, and finished in a marshy mess called Pegwell Bay. It's currently in the news for almost the first time, because my childhood backwater of Kent is to spawn a New Town, and there are plans to dam the Stour to supply its water. I am amazed that anyone believes such a little river can provide enough water to make it worthwhile. But when you are trying to water the driest corner of England, you take what you can get, apparently. And even the smallest river can fill the oceans if it flows for long enough.

Rivers so often define our world. Is there a better book about America than Huckleberry Finn's journey on the Mississippi? Is there a better way of seeing London than taking a boat down the Thames to Greenwich? Some of the greatest human adventures have been along rivers: up the Orinoco to find El Dorado, or the search for the source of the Nile. Millions of Indians keep bottles of Ganges water in their homes, like holy water. We romance on the Blue Danube and the Seine and fight over the Jordan and the rivers of Babylon.

And yet something disturbing has been happening. I only slowly became aware of it: just a news story here and there seeped through. But the maps in my atlas no longer seemed to accord with reality. Inland seas and lakes were disappearing. The old geography lesson about how rivers emerged from mountains, gathered water from tributaries, and finally disgorged their bloated flows into the oceans were now fiction. Many rivers were dying as they went on, not growing.

Because I am a journalist, I assembled a small file of press clippings. The Nile in Egypt, the Yellow River in China, the Indus in Pakistan, the Colorado and Rio Grande in the United States—all were reported to be trickling into the sand, sometimes hundreds of miles from the sea. Individually, these were interesting stories. Taken together, they seemed to me to be something more. Some kind of cataclysm was striking the world's rivers. And so began the idea for this book.

I soon learned more. Israel is draining the Jordan River into pipes before it reaches the country that bears its name. There has been drought on the Ganges, because India has sucked up the holy river's entire dry-season flow. The great Oxus, the Nile of Central Asia, was diverted into the desert, leaving the Aral Sea to dry out. This was a real shocker; the shoreline shown on most maps of what was once the world's fourth largest inland sea was hundreds of miles distant from the reality. Even the chalk streams of my childhood were disappearing.

The wells have been drying up, too. Half a century of pumping on the Great Plains of the United States has removed water that will take two thousand years of rain to replace. In India, farmers whose fathers lifted water from wells with a bucket now sink boreholes more than half a mile into the rocks— and still they find no water.

My book is a journey of discovery on the world's rivers: to find out why we face this crisis, what happens when great rivers die, where we could be headed —and how we can restore the rivers' health and our hydrological future.

Although it is mainly about rivers, it is also about how we use water: about the staggering amount that it takes to feed and clothe us, and about how the world trade in food, cotton, and much more is also a trade in "virtual water"—

the water it takes to grow those crops. And that implicates Western consumers directly in the emptying of many of the world's great rivers.

As I took this journey, I met men in overcoats huddled in drafty offices, trying to keep the Ruritanian badlands of Karakalpakstan from turning to desert. I marched with a man who used to run the world's largest dam-building organization and now campaigns to tear down dams. I drank with Chinese bureaucrats who say that droughts on the Yellow River will one day trigger a flood disaster on a par with the one half a century ago that they try not to talk about—it was manmade and killed almost a million people. I met Indian rainwater harvesters and the last man alive with the secrets of divining water from underground Cyprus.

Here too are the stories of Colonel Qaddafi's Great Manmade River, pumping the world's greatest reservoir of freshwater from beneath the Sahara; of the world's third tallest dam, which any day could be ripped apart by either an earthquake or a civil war; of the river that flows backward to feed 10 million people; of the extraordinary ancient water tunnels beneath Iran that could stretch to the moon three times over; of the British-built irrigation project that is 60 miles from the nearest water. And here is the secret truth of the world's biggest ever water-poisoning scandal.

I hope to answer some pressing questions. Can we fill the world's faucets without emptying its rivers? Do we need megaprojects to empty the Great Lakes into the American West, the Congo River into the Sahara, or the torrents of Siberia into the deserts of Central Asia? Or should we think small, catching the rain from our roofs and irrigating crops with bicycle inner tubes or plastic sheaths bought from ice cream salesmen?

Nothing, perhaps not even climate change, will matter more to humanity's future on this planet over the next century than the fate of our rivers. Plenty of explorers have sought the source of the world's great rivers. This is a journey to chart their deaths. But it is a hopeful journey nonetheless. I am an optimist. Water, after all, is the ultimate renewable resource.

I

When
   the rivers
run dry...

the crops fail

1

The Human Sponge

Few of us realize how much water it takes to get us through the day. On average, we drink no more than a gallon and a half of the stuff. Including water for washing and for flushing the toilet, we use only about 40 gallons each. In some countries suburban lawn sprinklers, swimming pools, and sundry outdoor uses can double that figure. Typical per capita water use in suburban Australia is about 90 gallons, and in the United States around 100 gallons. There are exceptions, though. One suburban household in Orange County, Florida, was billed for 4.1 million gallons in a single year, or more than 10,400 gallons a day. Nobody knows how they got through that much.

We can all save water in the home. But as laudable as it is to take a shower rather than a bath and turn off the faucet while brushing our teeth, we shouldn't get hold of the idea that regular domestic water use is what is really emptying the world's rivers. Manufacturing the goods that we fill our homes with consumes a certain amount, but that's not the real story either. It is only when we add in the water needed to grow what we eat and drink that the numbers really begin to soar.

Get your head around a few of these numbers, if you can. They are mind-boggling. It takes between 250 and 650 gallons of water to grow a pound of rice. That is more water than many households use in a week. For just a bag of rice. Keep going. It takes 130 gallons to grow a pound of wheat and 65 gallons for a pound of potatoes. And when you start feeding grain to livestock for animal products such as meat and milk, the numbers become yet more startling. It takes 3000 gallons to grow the feed for enough cow to make a quarter-pound hamburger, and between 500 and 1000 gallons for that cow to fill its udders with a quart of milk. Cheese? That takes about 650 gallons for a pound of cheddar or brie or camembert.

And if you think your shopping cart is getting a little bulky at this point, maybe you should leave that 1-pound box of sugar on the shelf. It took up to 400 gallons to produce. And the 1-pound jar of coffee tips the scales at 2650 gallons—or 10 tons—of water. Imagine taking *that* home from the store.

Turn these statistics into meal portions and you come up with more than 25 gallons for a portion of rice, 40 gallons for the bread in a sandwich or a serving of toast, 130 gallons for a two-egg omelet or a mixed salad, 265 gallons for a glass of milk, 400 gallons for an ice cream, 530 gallons for a pork chop, 800 gallons for a hamburger, and 1320 gallons for a small steak. And if you have a sweet tooth, so much the worse: every teaspoonful of sugar in your coffee requires 50 cups of water to grow. Which is a lot, but not as much as the 37 gallons of water (or 592 cups) needed to grow the coffee itself. Prefer alcohol? A glass of wine or beer with dinner requires another 66 gallons, and a glass of brandy afterward takes a staggering 530 gallons.

We are all used to reading detailed technical information about the nutritional content of most food. Maybe it is time that we were given some clues as to how much water it took to grow and process the food. As the world's rivers run dry, it matters.

I figure that as a typical meat-eating, beer-swilling, milk-guzzling Westerner, I consume as much as a hundred times my own weight in water every day. Hats off, then, to my vegetarian daughter, who gets by with about half that. It's time, surely, to go out and preach the gospel of water conservation. But don't buy one of those jokey T-shirts advertised on the Internet with slogans like "Save water, bathe with a friend." Good message, but you could fill roughly twenty-five bathtubs with the water needed to grow the 9 ounces of cotton needed to make the shirt. It gives a whole new meaning to the wet T-shirt contest.

Let's do the annual audit. I probably drink only about 265 gallons of water —that's one ton or 1.3 cubic yards—in a whole year. Around the home I probably use between 50 and 100 tons. But growing the crops to feed and clothe me for a year must take between 1500 and 2000 tons—more than half the contents of an Olympic-size swimming pool.

———

Where does all that water come from? In England, where I live, most home-grown crops are watered by rain. So the water is at least cheap. But remember that a lot of the food consumed in Britain, and all the cotton, is imported. And when the water to grow crops is collected from rivers or pumped from underground, as it is in much of the world, it is increasingly expensive, and its diversion to fields is increasingly likely to deprive someone else of water and to empty rivers and underground water reserves. And when the rivers are running low, it is ever more likely that the water simply will not be there to grow the crops at all.

The water "footprint" of Western countries on the rest of the world deserves to become a serious issue. Whenever you buy a T-shirt made of Pakistani cotton, eat Thai rice, or drink coffee from Central America, you are influencing the hydrology of those regions—taking a share of the Indus River, the Mekong River, or the Costa Rican rains. You may be helping rivers run dry.

Economists call the water involved in the growing and manufacture of products traded around the world "virtual water." In this terminology, every ton of wheat arriving at a dockside carries with it in virtual form the thousand tons of water needed to grow it. The global virtual-water trade is estimated to be around 800 million acre-feet a year, or twenty Nile Rivers. Of that, two thirds is in a huge range of crops, from grains to vegetable oil, sugar to cotton; a quarter is in meat and dairy products; and just a tenth is in industrial products. That means that nearly a tenth of all the water used in raising crops goes into the international virtual-water trade. This trade "moves water in volumes and over distances beyond the wildest imaginings of water engineers," says Tony Allan, of the School of Oriental and African Studies in London, who invented the term "virtual water."

The biggest net exporter of virtual water is the United States. It exports around a third of all the water it withdraws from the natural environment. Much of that is in grains, either directly or via meat. The United States is emptying critical underground water reserves, such as those beneath the High Plains, to grow grain for export. It also exports an amazing 80 million acre-feet of virtual water in beef. Other major exporters of virtual water include Canada (grain), Australia (cotton and sugar), Argentina (beef), and Thailand (rice).

Major importers of virtual water include Japan and the European Union. Few of these countries are short of water, so there are ethical questions about how much they should be doing this. But for other importers, virtual water is a vital lifeline. Iran, Egypt, and Algeria could starve without it; likewise water-stressed Jordan, which effectively imports between 80 and 90 percent of its water in the form of food. "The Middle East ran out of water some years ago. It is the first major region to do so in the history of the world," says Allan. He estimates that more water flows into the Middle East each year as a result of imports of virtual water than flows down the Nile.

While many nations relieve their water shortages by importing virtual water, some exacerbate their problems by exporting it. Israel and arid southern Spain both export water in tomatoes, Ethiopia in coffee. Mexico's virtual-water exports are emptying its largest water body, Lake Chapala, which is the main source of water for its second city, Guadalajara.

Many cotton-growing countries provide a vivid example of this perverse water trade. Cotton grows best in hot lands with year-round sun. Deserts, in other words. Old European colonies and protectorates such as Egypt, Sudan, and Pakistan still empty the Nile and the Indus for cotton-growing, as they did when Britain ruled and Lancashire cotton mills had to be supplied. When Russia transformed the deserts of Central Asia into a vast cotton plantation, it sowed the seeds of the destruction of the Aral Sea. Most of the missing water for the shriveling sea has in effect been exported over the past half-century in the form of virtual water that continues to clothe the Soviet Union.

Some analysts say that globally, the virtual-water trade significantly reduces water demand for growing crops. It enables farmers to grow crops where water requirements are less, they say. But this is mainly because the biggest trade in virtual water is the export of wheat and corn from temperate lands like the United States and Canada to hotter lands where the same crops would require more water. But for many other crops, such as cotton and sugar, the trade in virtual water looks like terribly bad business for the exporters.

Pakistan consumes more than 40 million acre-feet of water a year from the Indus River—almost a third of the river's total flow and enough to prevent any water from reaching the Arabian Sea—in order to grow cotton. How much sense does that make? And what logic is there in the United States pumping out the High Plains aquifer to add to a global grain glut? Whatever the virtues of the global trade in virtual water, the practice lies at the heart of some of the most intractable hydrological crises on the planet.

→ like Freedman article on oil/democracy → what will connection be the link b/w H2O + governance?

2

North America:
Crossing the Rio Grande

They serve a strong brew at the Alamo coffeehouse in Presidio, a small farming town near the U.S.-Mexican border. They need to. Times are tough, says Terry Bishop, looking up from his second mug. This land, next to the Rio Grande in Texas, has probably been continuously farmed for longer than anywhere in America, he says. Six hundred years, at least. It's been home to scalphunters and a penal colony; it's seen Comanche raids, Spanish missionaries, marauding Mexican revolutionaries, and a population boom during a recent amnesty for illegal aliens. All that time it has been farmed. But soon it will be back to sagebrush and salt cedar.

Climbing the levee by the river at the end of his last field, Bishop shows me the problem. The once mighty Rio Grande is now reduced to a sluggish brown trickle. In its middle stretches, the river often dries up entirely in the summer. All the water has been taken out by cities and farmers upstream. "The river's been disappearing since the fifties," says Bishop, who has farmed here since then. There hasn't been a flood worthy of the name since 1978. For 200 miles upstream of Presidio, there is no proper channel anymore, he says. They call it the forgotten river.

Bishop's land brings with it legal rights to 8000 acre-feet of water a year from the river—enough to flood his fields to a depth of more than three feet, enough to grow almost any crop he wants. But in recent years he has taken only a quarter of that. Even when he gets water, "it's too salty to grow anything much except alfalfa." But that's all academic now. Yields got so low, the farm went

Rio Grande

bust. Bishop leases some fields to tenants, but most of them are idle these days. The land is gradually returning to desert. And Bishop drinks a lot of coffee.

This is the way of things in Presidio. The town was once a major farming center. It used to ship in thousands of Mexican workers to harvest its crops. Bishop's farm alone once employed a thousand people. But that has all ended, and the unemployment rate among the town's permanent residents is almost 40 percent. About the only profitable business is desert tourism. An old silver mine a few miles up the road has been turned into a "ghost town," and a fort at Cibolo Creek is now an upmarket resort where Mick Jagger once stayed. Harvesting tourists, that's the game now, says Bishop.

On the map, the Rio Grande is the fifth longest river in North America and among the twenty longest in the world. Its main stem stretches almost 2000 miles, from the snowfields of the Colorado Rockies to the Gulf of Mexico via New Mexico and Texas. It drains a tenth of the continental United States and more than two fifths of Mexico. The hub of human exploitation of the Rio Grande is the Elephant Butte Reservoir, just upstream of El Paso, Texas. It was built in 1915 and changed the river forever. The wild, untamed flow, which obliterated villages and once rode right through downtown Albuquerque, was ended for good, and the river's waters were corralled for irrigation.

Today, Elephant Butte and its downstream sister, the Caballo, all but empty the river to supply El Paso and nearby farmers. Downstream the river is partly renewed with water from two tributaries, the Pecos, out of Texas, and the Rio Conchos, which comes in from Mexico and joins the main stem at Presidio, right by Bishop's farm. But this new water doesn't last long before being taken out to fill reservoirs supplying farms in the lower basin. More than 9 million people in the basin rely on the Rio Grande's waters. But it is the farmers who make most use of it. Four fifths of the water in the river is taken for irrigation —most of it to grow two of the thirstiest crops in the world, cotton and alfalfa (a grain that is fed to cattle). And the wastage is huge. Only about 40 percent of the water reaches the crops, and evaporation in the hot sun takes more than 6 feet of water a year from the reservoirs—a total of around 245,000 acre-feet from Elephant Butte alone.

Usually a trickle of water gets through to the sea. But since the mid-1990s, a decade during which drought gripped the basin, the flow has been at record lows. It should have come as no surprise when, on February 8, 2001, Cameron

County Agent Tony Reisinger took a photograph of the mouth of the Rio Grande in the delta at Boca Chica and the flow had ceased. A sandbar 325 feet wide had completely blocked off the river from the Gulf of Mexico. The bar lasted for five months before summer flows washed it away. And for much of the next two years it returned. You could drive a car across the beach between the United States and Mexico. Though a couple of storms raised the flow in 2004, nobody doubts that the river is in serious trouble.

My exploration of the strange death of the Rio Grande started in El Paso at the Chamizal National Memorial, which commemorates a treaty that fixed the boundary between El Paso and its Mexican twin city, Juarez, by forcing the meandering river to pass down an unchanging concrete canal. This brutal carve-up may have underlined the river's geographical importance, but it hardly accorded it respect. Today the river is virtually invisible from the memorial behind a high chain-link fence designed to keep out illegal Mexican immigrants. Only up on the ugly, heavily guarded border bridge can you see it—a fetid trickle in an absurdly wide concrete canal flanked by a six-lane highway and a container dump. There is so little flow that as I watched, the wind ripping upstream was blowing the water back toward its source in distant Colorado.

El Paso is in hydrological trouble. With the river now trickling through the town virtually empty and upstream reservoirs scarcely any fuller, the *El Paso Times* regularly alerts readers to the days when they can use public water on lawns and the days they can't. Jittery suburbanites are repairing old wells in the hope of capturing some private water from beneath their land. And in the unplanned shantytown *colonias* where Mexicans usually end up after crossing the river, thousands of people live without access to piped water at all—and that is a shock to find in the United States, even in the desert.

Across the border in Juarez, things are worse, of course. People there are so short of water that sewage effluent and salty underground water have become major resources. I visited a gleaming new plant that treats half the city's sewage and sends the cleaned-up effluent 25 miles downstream to irrigate crops. And I went to Anapra, one of the city's more notorious *colonias*, where migrant Raphael Valarez told of his delight that he could now walk down the street to collect water from a new desalination plant. He had so much water his young daughter could paddle in a big washtub.

Meanwhile, El Paso is buying up properties from local farmers to bag their rights to underground water reserves. Its "water ranches" are dotted all along the highway to Presidio. Tiny cattle towns like Valentine and Dell City could in future provide half of El Paso's water. But water ranching is only a temporary solution. The major aquifer here—which underlies 15 million acres of Texas, New Mexico, and the Mexican Chihuahua desert—is rapidly running dry, because there isn't enough rain to replace what is pumped out.

As we drove east, Mary Kelly, a leading environmentalist campaigning to save the river, said she feared that pumping out this underground water will trigger the final and irreversible desiccation of the landscape. "Most of the recharge for the aquifer comes from seepage from the canals supplying the farms —which ultimately means it comes from the Rio Grande," she said. Now the government wants to line the canals to save water, so the seepage will stop and the aquifer will die even faster. "That is not a good plan," she added. But as I discovered on my subsequent journey around the world, it is exactly the kind of thing that happens when rivers run dry.

Downstream of El Paso and Juarez, the "forgotten river" is worthy of the name—a dribble of sewage effluent that disappears into remote scrub most of the way to Presidio. An invasion of salt cedar has taken over the channel, sucking up what liquid remains. Salt cedar, also known as tamarisk, is a nasty and extremely tough shrub, able to withstand fire and drought, flood and searing desert heat. A single plant can drink more than 265 gallons of water a day.

Hydrologically speaking, the Rio Grande pretty much ends here. The river that rolls out of the Colorado snowpack and once ransacked Albuquerque dies now amid the salt cedars and saline pools of the forgotten river. The only water for the channel comes from Mexico. Back in Presidio, standing on the levee on Terry Bishop's land, I could see the empty Rio Grande bed being filled from the south by the Rio Conchos. For all practical purposes, the Rio Grande downstream of Presidio was in fact the Rio Conchos.

Beyond Presidio the river winds through dramatic canyons in the Big Bend National Park. But the flow is small and muddy. "We get about a sixth of the historical flow here," Dave Elkovitz, of the National Park Service, told me. A couple of weeks before my visit, the river dried up in the park for the first time in more than fifty years. Stagnant pools of water evaporated, leaving dry gravel beds and thousands of dead catfish. Starved of food, a troop of black bears headed back to Mexico. Border patrols in jeeps drove up the riverbed, pursuing those who originally generated the term "wetbacks" but who could now simply walk across the border without wetting their feet.

"We have treaties for the river," said Elkovitz. "But they allocate more water than actually exists. What good is that?" The main treaty, signed back in 1944, when northern Mexico was much emptier than it is today, certainly sounded one-sided to me. It might have been fine half a century ago, when the rains could be relied on, but it puts a bullet to the head of Mexican farmers in modern times of drought. The treaty requires that one third of the water flowing into the Rio Grande from six Mexican tributaries, much the largest of which is the Rio Conchos, is allocated to the United States. Worse, however low the tributaries flow and however little rain falls in the deserts of northern Mexico, the Mexicans are still required to deliver annually for U.S. use a minimum of 350,000 acre-feet, averaged over five years. The United States can dry up the Rio Grande at El Paso as much as it wants, but come hell or low water, the Mexicans have to deliver that quota.

That, at any rate, is the theory. Drought and the demands of its own farmers left Mexico more than four years in arrears by 2003. Some good rains south of the border in early 2004 relieved things a little, and Mexico reduced its water debt to three years. But by late 2004, Texas farmers were back on the warpath, demanding reparations of $1 billion for crop losses over the previous decade. "We want to be good neighbors, but it's hard to do when you are losing money," said citrus farmer Jimmie Steidinger, who estimated he had personally lost a quarter million dollars. A debt is a debt, even in water flowing down a river.

———

Thunder clapped in the canyons and it started to rain as I headed over the border to Presidio's twin town, Ojinaga, on the Rio Conchos. Long before the arrival of Europeans, native Americans lived along this river, hunting the animals that congregated beside its waters. Early Spanish explorers came north up the Conchos before moving on into Texas and New Mexico. A century ago, Ojinaga was the headquarters for Pancho Villa's Mexican revolution.

The Ojinaga Irrigation District has been a source of prosperity here for a generation, taking water from the Rio Conchos. The district covers 22,000 acres, growing cotton and corn in the Chihuahua desert. But here, as in Presidio, farming seems to be on its last legs. Under the shade of cottonwood trees on his small farm, I met sixty-eight-year-old Rito Guerrero. He came down from the Sierra hills to farm here when he was young and the irrigation canals were new. "Once, the water came right up from the river. All my land would be watered," he told me. He still gets water sometimes from the canals that run right past his gate, but a combination of water shortages and falling prices for crops is forcing him out. "It's six years since I planted corn," he said. "I tried watermelons two years ago, but I lost everything. Cotton had a good market, but there isn't enough water. I could be living in the U.S.A. with my son." One day soon he will probably join the diaspora.

Farmer cum teacher Humberto Lujan took me around the irrigation district. Most of the canals were dry and their sluice gates closed and rusting. Three quarters of the fields were abandoned. The sagebrush was returning. "Virtually nothing comes down the river anymore," said Lujan as we peered into the town's empty reservoir and spotted the invasion of salt cedar along the riverbank. The town has lost a quarter of its population. "The young have gone north to the States, and the old are selling up," he explained.

As Lujan sees it, northern Mexico has been living an impossible dream for too long. For thirty years people flocked here from across the country to grow water-guzzling crops. Before they dug the irrigation canals, he remembers, "There was a traditional irrigation system here—a few small dams on the streams made with the branches of trees. It was a lot more efficient. It had less water but served more acres." They grew onions and melons and corn for the local markets. "Every *campesino* had chiles and tomatoes and a few cows for milk." Then they built the dams and canals to grow cotton. They ripped up all the existing vegetation and destroyed the soil, which became salty. Then the water ran out. "Before, we had irrigation here and good soils," Lujan told me. "But we have ended up creating desert."

Northern Mexico is littered with irrigation areas living on borrowed time and the water they owe to Texas. I took the bus south through the desert to the Delicias Irrigation District, which stretches for more than 60 miles east of the town of Chihuahua. It is the largest irrigation district on the Rio Conchos. Its glory days were in the 1960s and 1970s, when farmers became rich growing cotton, which they called "white gold." Today it is the cornerstone of a big dairy industry. The cattle feed on alfalfa grown on its irrigated fields. This is madness. Alfalfa is an extremely thirsty crop. Out here in the desert, where water shortages are critical, farmers use more than 2000 gallons of irrigation water to produce a gallon of milk.

Most of the fields are fed from the river's Boquilla and Madero dams. The Madero Dam is decorated with two statues, one carrying corn and the other cotton. It has a recreation area with water slides and advertises itself as a fishing resort. The initial effect was impressive but misleading. The reservoir was at a third of capacity. Beneath a bridge over the 1300-foot-wide riverbed, the stream of water was only about 6 feet wide. The reservoir was shutting off the river's entire flow. It turns out that of the district's 10,000 farms, only about a third can be sure of getting any water most years. And none now get water in winter as they once did. "Some areas have not been irrigated for ten years," said the district engineer Ezequil Bueno Torres. The farmers complain that the government gives away all their water to Texans. One, Rogelia Bejarano, was running in the upcoming local elections. He had a simple campaign pitch: "I won't give away the water; I'll give it to the farmers here."

Meanwhile the district is engaged in a big modernization process to save water. But its efforts are likely to prove just one more step in an escalating hydrological tragedy. Engineers like Torres are lining canals to prevent water from seeping through their porous bottoms, and they are installing perforated rubber piping to get water right to the plant roots rather than flooding fields. The $130 million modernization is paid for mostly by the U.S. government, and the aim is to save 285,000 acre-feet of water annually. That volume is close to the amount that Mexico should be sending down the Rio Conchos into the Rio Grande each year. "The Americans will get what we save," said officials at the dam.

It sounds like a win-win situation, and almost everyone I spoke to considered it a great idea. The Mexican farmers, I was told, would get greater water security, while the United States would get its missing water. But there is a fallacy here, just as there is a fallacy in lining the canals around El Paso. The modernization won't actually make any more water, and most of the savings are not real savings. As the canal water has failed in recent years, many farmers have come to rely on pumping underground water. That already accounts for about a fifth of all the irrigation in Delicias. Farmers have dug holes up to 130 feet wide that fill with underground water. One of the pioneers, Marcial Marquez, said he could pump from the holes dug on his land every day without their emptying. But most of the underground water in the wells comes from seepage from the canals and fields—the supposedly wasted water. If the seepage is cut, Marquez's wells will dry up.

The local irrigation engineers, and the people funding their work, seem unconcerned by this prospect. But the tragedy is that to meet their immediate obligations to deliver water to Texas farmers, the Mexicans are imperiling the long-term future of their underground water reserves. The win-win could quickly turn into a lose-lose as the ratchet emptying the Rio Grande is given another turn, this time in the name of efficiency.

3

Riding the Water Cycle

Earth is the water planet. It contains an unimaginable 1.1 quadrillion acre-feet of the stuff. But more than 97 percent of it is seawater, which we cannot drink and cannot, except in very local circumstances, afford to purify. "Water, water everywhere, / Nor any drop to drink," as Coleridge's Ancient Mariner put it. Of the remaining 28 trillion acre-feet of freshwater on and near the planet's surface, two thirds is locked up in ice caps and glaciers and one third, about 9.7 trillion acre-feet, is in liquid form. The greater part of it is in the pores of rocks. These reservoirs of underground water, known as aquifers, vary hugely in their accessibility and drinkability. But the water is there, beneath our feet.

The remaining smidgeon of the world's liquid freshwater—we are now down to a mere 162 billion acre-feet or so—is above ground. The biggest volumes, around 71 billion acre-feet, are in lakes, and there are probably another 71 billion acre-feet in soils and permafrost. Next comes atmospheric water vapor, which contains another 10.5 billion acre-feet, or about a thousandth of 1 percent of all the water on earth. After that come swamps and wetlands at 9 billion acre-feet; rivers, which contain around 1.6 billion acre-feet at any one time; and living organisms, from rainforests to you and me, with about 800 million acre-feet.

But this is a static picture: where the water is at any one moment in time. And water is not a static resource. It is constantly on the move, through soils and into ancient geological formations, down rivers, into the ocean depths, freezing and melting, evaporating into the air and forming clouds to fall again as rain. If we are concerned about how much water is available to us in a form that we can keep on using, the static figures don't tell us much. It is the dimensions of these movements, known as the water cycle, that are critical.

Take aquifers. The crude statistics suggest that they should easily be our number-one resource. And yes, the volume of water down there is huge, dependable, and accessible, ready to be pumped to the surface. With current technology, perhaps a tenth of the known resource could be extracted. Sink a well and away you go. But as everyone knows, wells run dry. Pumping out an aquifer will empty it forever—unless that aquifer is being refilled. So a more useful measure of the water available to us underground, year in and year out, is how much of that water is replenished by rain.

Looked at this way, underground water shrinks in importance. By some perversity of nature, many of the biggest aquifers are beneath deserts and get virtually no recharge because there is no rain. The largest quantities sit in the pores of sandstone rocks beneath the Sahara and the Arabian peninsula. Other big reserves are beneath the Australian outback and the arid High Plains of the American Midwest. We can and do pump this water, but as we do, the water level falls, pumping costs rise, and quite often water quality deteriorates. Probably about a tenth of 1 percent of the freshwater in the world's aquifers is replaced each year by rains. So while the amount of time that water spends in underground rocks varies from a few months to millions of years, on average it is getting on for a thousand years.

Water in the other big reservoirs, the oceans and ice caps, has a similar long residence time. But if most of the water in what we might call the "slow" water cycle moves at a snail's pace, water in other parts of the water cycle—generally the parts with smaller volumes at any one time—moves around much more quickly. There is a "fast" water cycle, too—a constant rapid flow of water evaporating from land and sea, forming water vapor and then clouds, raining to the earth and then evaporating again.

Of course, some water from the fast cycle disappears into the slow cycle. Rain percolates into rocks or gets frozen on glaciers or trapped in inland seas or tied up in plants and animals. And some water moves from the slow to the fast cycle—in meltwater from glaciers, or bubbling from rocks in springs, for instance. But in a year, about 400 billion acre-feet of water flows through the fast water cycle. In practice, there is much less water; it just circulates several times. But if this entire annual cycle were collected together, it would cover the entire earth's surface to a depth of almost three feet.

Because most of the planet is covered in ocean, the larger part of this fast cycle is again of not much use to us. It comprises water evaporating from the sea and falling back onto the sea in rain. For the rest, we are beginning to home in on the part of the cycle from which we get most of our water.

First there is the water that falls as rain onto the land and then evaporates from the land again. This involves about 49 billion acre-feet each year. Some evaporates from soil; some is taken up by plants as they grow and is then released from leaves in a process called transpiration; some evaporates from bodies of water, including manmade reservoirs. Second, there is a constant flow of water from the land, where rainfall exceeds evaporation, to the sea, where evaporation exceeds rainfall. Some of this flows directly off the land, some shoots down gullies that form when it rains, and some flows down permanent rivers. Most of the water we use around the world today comes from this part of the water cycle. About 32 billion acre-feet of water makes the journey from the land to the sea every year. Or it did before we started diverting it.

So how much of this runoff do we use? Of those 32 billion acre-feet a year, much hurtles off the land in occasional floods or flows away from permanent rivers. Of the rest, hydrologists estimate the maximum that might reasonably be caught and used by humans employing current technology is 11 billion acre-feet. But nature has played one more trick on us. Many of the world's greatest rivers are in regions where few people can or want to live. The three rivers with the biggest flows—the Amazon, the Congo, and the Orinoco—all pass through inhospitable jungle for most of their journey from headwaters to the sea. Those three alone carry almost a quarter of the water we have to survive on. And two more of the top ten—the Lena and the Yenisei, in Siberia—run mostly through Arctic wastes. A tenth of the world's river waters flow into the Arctic. Take out these and we are left with around 7 billion acre-feet of river water for our needs.

That still amounts to about 370,000 gallons a year for every citizen on the planet. Not bad, but I calculated my own annual water use at between 400,000 and 530,000 gallons a year. I imagine most of the world would like to live as well as I do. So we have a problem.

———

Water is heavy stuff. It's not easy to move around unless gravity is with you, as anyone who has carried a bucket of the stuff any distance will agree. So all these global numbers can still give a misleading impression of how much water we can really get our hands on. The water business, like the property business, comes down to a matter of "location, location, location." So where is the water?

Just six countries have half of the world's total renewable freshwater supply on their territory—Brazil, Russia, Canada, Indonesia, China, and Colombia. But some smaller countries do even better if we measure the amount of water available per capita of population. Greenland's 60,000 citizens have more water than anyone else. Each of them could make use of 8 million gallons of it every day without having to defrost a single piece of ice. But having no crops to irrigate, they need little. Alaskans have available 1 million gallons a day each. Also weighing in above 130,000 gallons each a day are the citizens of the Congo, Iceland, and the three neighboring South American rainforest states of Guyana, Suriname, and French Guiana.

Meanwhile, people in the driest countries have by far the greatest need—to irrigate crops as well as to quench their thirst. The Palestinian desert enclave of the Gaza Strip is the most water-starved political unit on earth, with just 37 gallons of brackish underground water a day for each inhabitant. Others at the bottom of the hydrological heap include small desert states such as Kuwait and the United Arab Emirates, and island states like the Bahamas, the Maldives, and Malta.

Even at the continental level there are clear haves and have-nots. Europe and North America are well served—especially considering that irrigation demands are mostly fairly low—though both continents generally have a surplus in the north and shortages in the south. Australia is the driest continent but has a small population. Asia, by contrast, has almost two thirds of the world's population but only a third of its runoff, 80 percent of which is concentrated in the short monsoon season.

Africa's continent-wide allocation isn't too bad, but a third of its runoff goes down a single river, the Congo. South America has only 5 percent of the world's population but glories in three of the world's top ten rivers—the Amazon, the Orinoco, and the Parana. Fully a quarter of the world's runoff occurs here, but little of this water is near people. The vast and wet Amazon basin, with 15 percent of the entire world's runoff, has just 0.4 percent of the world's population.

Even within countries, water can be inconveniently distributed in time and place. Most of India gets all its rain in a hundred hours during a hundred days. Parts of Ethiopia suffer perennially from drought and famine, even though 84 percent of the flow of the Nile River begins within the country's borders. Canada has 90 percent of its water where 10 percent of its people are, and vice versa. If northern China were a separate country, it would be one of the most water-stressed in the world. All this has come to matter a very great deal because of two changes over the past half-century: the soaring world population, and the manner in which we have gone about trying to feed that population.

Around forty years ago, the world was gripped by Malthusian nightmares of mass starvation. Women were having an average of more than five children each. The global population was set to double in a generation. Billions, it was said, would die of famine. In his influential book *The Population Bomb,* the biologist Paul Ehrlich declared that "in the 1970s hundreds of millions of people are going to starve to death . . . the battle to feed all of humanity is over." And as famines spread through Africa in the early 1970s, he was not alone in holding such apocalyptic views. A crude computer model of ecological Armageddon called Limits to Growth became a worldwide bestseller.

But it didn't happen that way. The global population has since doubled, but scientists defied the doomsayers by breeding new high-yielding varieties of rice, wheat, and corn that kept the granaries full. And Henry Kissinger was proved right: he got up at a World Food Conference in 1974 and declared that "for the first time, we may have the technical capacity to free mankind from the scourge of hunger." He somewhat overestimated the world's ability to get food to those who needed it, so his follow-up—that "within a decade no child will go to bed hungry"—was sadly wide of the mark. Africa in particularly has missed out. But, averaged globally, the "green revolution" has ever since kept growth in food production ahead of growth in population.

It is an impressive achievement. But those new high-yielding crop varieties needed water, huge amounts of it. So the world embarked on a vast investment program, first in dams and then in irrigation canals to deliver that water to fields. With a typical bill for supplying water from rivers to fields at between $400 and $4000 per acre, the financial cost has been huge. But today poor developing countries like India, China, and Pakistan have three quarters of all the world's irrigated farmland. And many of them have defied the doomsayers by moving from famine to self-sufficiency in their staple grains.

The problem now is escalating shortages of water. Today some 70 percent of all the water abstracted from rivers and underground reserves is being spread onto the 670 million acres of irrigated land that grows a third of the world's food. This massive global undertaking has kept the world's granaries full, but it has emptied the rivers. Back in the 1960s and 1970s, neither the green-revolution scientists nor the doomsayers fully appreciated that while the new crops were indeed very efficient at delivering more crop per acre, they were often extremely inefficient when measured against water use. They often produced less "crop per drop" than the varieties they replaced.

As a result, the United Nations' Food and Agriculture Organization says that on at least a third of the world's fields today, "water rather than land is the binding constraint" on production. Perhaps the most telling statistic of all is this: the world grows twice as much food as it did a generation ago, but it abstracts three times more water from rivers and underground aquifers to do it.

In arid countries such as Egypt, Mexico, Pakistan, and Australia and across Central Asia, 90 percent or more of the water abstracted from the environment is for irrigation. In the green-revolution countries, water consumption per capita is several times that of European countries. Pakistan abstracts five times more water per person than Ireland does, Egypt five times more than Britain, and Mexico five times more than Denmark. No wonder their rivers are emptying.

One straw in the wind: the Mexican valley that once described itself as "the home of the green revolution," because its farmers pioneered use of new high-yielding wheat varieties half a century ago, has run out of water. Reservoirs on the Yaqui River in the Sonoran Desert of northwestern Mexico have dried up since the mid-1990s as farmers' demand exceeded supply in the rivers. Today fields are being abandoned and farmers are leaving the valley.

The Yaqui Valley is far from alone. Since the 1970s, Egypt, for instance, has had to import growing amounts of food because there is not enough water to sustain the new high-yielding crops. And other countries are living on borrowed time. In some cases the water is running out. Thus, a quarter of India's crops are being grown using nonrenewable underground water. In others, salt from the irrigation water is invading the fields and rendering large areas sterile and useless. Each year 25 million acres of fields around the world are lost to its toxic embrace. Projects that initially greened the desert are now creating desert.

It is dangerous to predict doom when it didn't happen last time. Science may triumph again. But without a second agricultural revolution that targets water, a "blue revolution," then the gains of the past generation could be wiped out as rivers run dry, underground water reserves are exhausted, and fields are caked in salt.

green revolution

high yielding crops

detriment of technology

HIGH water use

BLUE REVOLUTION

need.

4

profits + perils

Pakistan:
The Unhappy Valley

If any country sums up the profits and perils of the overuse of water for irrigation, it is Pakistan. The saga began with British colonial engineers. When they first surveyed the Indus Valley in the nineteenth century, they saw with delight a near-replica of the Nile yet three times the size. The snow melting each spring in the Himalayas provided a vast volume of water all down the Indus to its delta on the Arabian Sea. The river could, they decided, be engineered to irrigate its wide desert plain: a land of fertile soils but no rain.

Like the land by the Nile, the Indus Valley had a long indigenous tradition of irrigating fields along the banks of the river. The Harappa civilization had prospered in this way five thousand years before, diverting the summer monsoon flows through the river's banks and onto their fields. The practice of raising temporary earth dams to force the river and its tributaries to flood bankside fields continued right into the nineteenth century, when the British decided that they could do things better. They wanted to trap the summer floodwaters permanently behind massive barrages to provide year-round irrigation that could sustain two or even three annual crops. The British were dreaming of imperial splendor, but they were driven at least as much by the prospect of financial gain. They saw full granaries across the Indus Valley as a source of tax revenues for the Treasury in London.

In the words of the imperial historian James Morris, they would "create a new country out of a gloomy wasteland," turning a land peopled only by "disputatious nomads who made a living by cattle-thieving" into something more like home. The first and greatest creation was to transform the Sikh state of Punjab. To do this, after sending in the soldiers to subjugate the locals, the engineers built a succession of barrages across the great river and its tributaries where they emerged from the Himalayas and extended canals for hundreds of miles through the wide, flat plain. In 1892 they completed the largest of the canals, Lower Chenab Canal, which carried six times the flow of the Thames River back home and irrigated an area the size of Yorkshire.

They were remaking the social landscape, too. The British threw the by then understandably disputatious nomads off their land and marched in farmers from central India. They dotted the plain with villages and established market towns at regular intervals along the canals. At the heart of the new Punjab they laid out a city in the pattern of the Union Jack, naming it Lyallpur after the local governor. (After independence it was renamed Faisalabad). There they built what remains to this day the largest agricultural university in Asia.

No Stalinist social engineer could have ordered things more tidily, though the British added their own element. The city's new geography reflected clear ethnic divisions, with Sikhs in one area, Muslims in another, and Hindus in a third, but also British class sensibilities, with the imported settlers divided between peasants, yeomen, and gentry. The last group, the officer class, got the biggest plots and were expected to be natural leaders.

By its own lights, the project was a huge success, turning the backward province into one of the most productive in the Indian empire. But there was still water in the river, flowing past the barrages and through the desert of Sindh Province to the rich green delta and the Arabian Sea. By now the British had a new market they wanted to supply: the cotton mills in Lancashire. So in 1932 they finally, and probably somewhat reluctantly, trekked south into the Sindh desert, which they had called "the unhappy valley," and began building new hydraulic projects. At their heart was the Sukkur Barrage, which strode across the Indus in sixty-six arches and diverted most of the river's remaining flow into the desert, where the British established what amounted to one giant cotton farm.

British engineering endeavors in the unhappy valley eventually produced what remains the largest unbroken irrigated area on the planet. Even California cannot compete. And it was just one part of Britain's grand plan to irrigate its great Indian empire, the jewel in Queen Victoria's crown. Engineers such as

Sir Arthur Cotton—who had sailed for India as a callow youth of sixteen, intent on making his way in the empire, and became the architect of the greatest irrigation network the world had ever seen—thought nothing of damming rivers far bigger than any they knew back home in England.

With the help of countless "coolies" carrying soil in baskets on their backs, they turned the muddy flats of the Cauvery delta in Tamil Nadu into a vast rice paddy covering some 1000 square miles. They created a dam near Bombay that was, they claimed, the largest piece of masonry erected since classical times. They dug canals the length of the great northern plain of the Ganges, channeling most of the river's winter flow to fields. "Great tracts of country once desert and without inhabitants are now filled with people," declared John Colvin, the governor of the northwest provinces, one morning as he opened the largest brick aqueduct in the world, over the Solani River at Roorkee. In these great works, says the technology historian Daniel Headrick, "they laid the foundations of modern hydraulic engineering."

The British-built railroads of India are more famous, but the empire spent far more on digging canals than on laying tracks. Britain ended its rule in 1947 with more than 60,000 miles of canals carrying water the length of the subcontinent, and with more irrigated land than there is land of all sorts in the whole of England. As the sun went down on the British Empire in India, one in every eight formally irrigated fields in the world was in India. And the greatest irrigated expanse of them all was on the Indus.

———

But the engineers were not finished. The newly independent Pakistan, funded by the World Bank and other agencies set up after the Second World War, strove to enhance its control over the Indus yet further by building a series of giant dams on the upper reaches of the river, in the foothills of the Himalayas. One such was the Tarbela, which, when it was completed in 1974, was the world's largest. These upstream dams were intended both to generate electricity and to store water in wet years, thus drought-proofing the country in dry years. And Pakistan has since built two more barrages in Sindh to divert more Indus water into the desert and to increase its irrigated land further—to around 62,000 square miles in all.

Today no other country but Egypt is as dependent on one river as Pakistan is on the Indus. Just as the Greek geographer Herodotus two thousand years ago called Egypt "the gift of the Nile," so modern Pakistan, a country of 140 million people, is "the gift of the Indus." Without the river, most of the country would be desert. Besides watering about 90 percent of its crops, the river produces nearly half of its electricity. Thanks to irrigation, the country is one of the world's leading exporters of cotton and manufacturers of textiles. The Pakistani Water and Power Development Authority is the country's largest civilian organization.

But there are escalating problems, made more critical by Pakistan's huge dependence on the river. The hydraulic regime created over 150 years contains the seeds of its own downfall, and they are multiplying fast. The barrages in Sindh and Punjab have diverted so much water onto the flat Indus plains that the soils are becoming waterlogged. And salt brought down by the river is accumulating in the fields, poisoning crops.

By one calculation, the river delivers 24 million tons of salt onto the plain each year, but it removes only 12 million tons to the Arabian Sea. The rest, close to a half-ton a year for every irrigated acre, stays on the fields. Farmers have responded in the only way they can, by pouring on every available drop of water to wash away the salt before they plant crops. But with water in short supply, that is increasingly difficult to achieve. When it cannot be done, the salt forms a white crust on the surface of the soil and the fields die.

In recent years farmers have been abandoning salt-encrusted land at a rate of 100,000 acres a year. A tenth of the fields have been lost so far; of the remainder, a fifth are badly waterlogged and a quarter produce only meager crops. The situation is worst in Sindh Province, where four out of five fields are waterlogged. In some parts, more than half the land is barren, with a few clumps of grass poking through the white crust of salt. Hugely expensive programs to install drains across Sindh in an effort to empty fields of the salt have been only partially successful.

Meanwhile, the country's population is soaring. It has quadrupled since independence and is projected to reach 250 million by 2025. These people all need feeding. Nationally, despite the lost land, food output has so far kept pace with the rising population, thanks to new high-yielding crop varieties. But this seems unlikely to continue. As salt builds up, it takes more and more water to enable fields to produce. And there is no more water to be had in the country.

Typically, 57 million acre-feet of the Indus's water is used to grow rice and a further 40.5 million acre-feet each to grow wheat and cotton. That is a total of 138 million acre-feet—out of an average flow of 146 million acre-feet. And in the majority of recent years, the average has not been reached. The gloomy prognosis is that as Pakistan's population continues to soar, the country will have less and less land and water capable of producing the crops with which to sustain itself.

Politicians in the capital, Islamabad, want to build more dams to secure more water. At the top of the list is the Kalabagh Dam, downstream of the Tarbela Dam. It would, they say, extend irrigated farming into the unplowed west of the province. But there are even grander plans. They want to flood the city of Skardu, the capital of an area known as Baltistan, or Little Tibet, to create a huge dam in the distant Karakum Mountains on the northwest frontier. This would store the waters that occasionally rush out of the mountains in "superfloods," which other dams cannot contain.

But downstream in Sindh, farmers believe the schemes will simply capture more water for Punjabi farmers and leave their own canals empty. Punjab and Sindh often seem on the brink of a water war. Back in 1998, the Sindh leader Benazir Bhutto led tens of thousands of farmers in demonstrations against the Kalabagh project. Three years later, two people died in riots during a political rally to protest water shortages and bombs exploded in Karachi during strikes against water shortages. "The word Kalabagh has become synonymous with Punjabi chauvinism," says a local commentator, Ayaz Amir. If there is ever a war between the provinces, Kalabagh is likely to be the spark.

Whatever the politics, and however the water is shared, it is abundantly clear that the Indus is in deep trouble. In the first years of the twenty-first century, the river was largely dry for its final few hundred miles to the sea. The Indus delta was once a rich expanse of mangrove swamps stretching for hundreds of miles, interspersed with pastures and creeks where river dolphins swam and fish spawned. Even half a century ago, it got water for several months a year. But years of meager flows since have killed half the mangrove swamps. Many of the fish have vanished. One economic assessment puts the loss of fish alone at $20 million a year. Even drinking water has become scarce.

With little silt or freshwater reaching the delta, the sea is advancing inland. At least a million acres of mangroves and farmland have disappeared beneath the waves. Life is becoming intolerable, says Mohammed Tahir Qureshi, an ecologist with the World Conservation Union. "More than three quarters of the local population depend on these products for their livelihoods, so there has been a mass migration out of the area," he explains.

As the irrigation systems fail and the natural resources of the Indus and its delta expire, the social cohesion that the British once so deliberately sought to create in Pakistan is also fast breaking down. The farms of Sindh could once employ everyone. But today gangs of unemployed bandits stalk the country-side, exacting protection money from those farmers still in business. And millions of people have left the fields and moved to Karachi. During the 1990s, this former colonial port near the mouth of the Indus became the fastest-growing city on earth. Swelled by disgruntled environmental refugees, it now has more than 10 million inhabitants. Most live in the city's vast lawless slums, which are a breeding ground for Al Qaeda and other terrorist groups.

Irrigation has made, but it has also destroyed. Sindh is once again, as the British called it before they marched south to remake its hydrology, the unhappy valley.

[handwritten margin note: ✱ link between environmental degredation + Al Qaeda ↓ eco-refugees / extreme poverty ⇒ slums]

II

When
the rivers
run dry . . .

we mine our children's water

5

India:
A Colossal Anarchy

[handwritten notes: green revolution centered on high yield crops + irrigation; irrigation → river canals / underground reserves]

A generation ago, Indians were starving. Their grain stores were empty and the doomsayers were gathering with the vultures. Today the stores are full and the fear of famine has receded, thanks to a green revolution built largely on irrigation. India, like many of its Asian neighbors, has kept bellies full while its population has doubled. This has come about as a result of a new generation of high-yielding varieties of staple crops such as rice, wheat, and corn. The key to the success of these supercrops has been abundant supplies of irrigation water. Most other Asian countries have achieved agricultural self-sufficiency by emptying their rivers into irrigation canals, but India's green revolution is increasingly being watered by plundering the country's underground water. More than 21 million farmers now tap underground reserves to water their fields.

This self-reliance is born partly of the administrative failure and technical inefficiency of many surface irrigation projects built by India in the early years of the green revolution. The prime minister then, Rajiv Gandhi, effectively wrote their obituary in 1986 when he complained that only a quarter of the 246 large surface irrigation projects begun in the previous thirty years had been completed. "Perhaps we can safely say that almost no benefit has come to the people from these projects," he said. "For sixteen years, we have poured money out. The people have got nothing back: no irrigation, no water, no increase in production, and no help in their daily life."

He was exaggerating. But his diagnosis that "the drama of harnessing a major river may be more exciting than the prosaic task of getting a steady trickle of water to a parched hectare" was correct. Indian engineers liked building dams and canals more than creating the fiddly works needed to get the expensively collected water to farms and fields or to insure that that water was used efficiently. And their political masters liked grand openings of large structures—often festooned with plaques with their names on them—more than the actual business of feeding their people.

But the failure of surface irrigation projects was not just about administrative indolence. India's rivers simply do not contain enough water to sustain the demands being made on them, so millions of farmers have taken things into their own hands. They are hiring drilling rigs and buying $600 electric pumps to mine water that has lain undisturbed beneath their land for millennia. Most of the water goes onto the fields, though some enters India's burgeoning private water market.

India's farmers have spent in the region of $12 billion on pumps and boreholes in the past two decades. In many ways they should be congratulated for their enterprise and investment. They have kept India fed. And there is plenty of evidence that farmers who tap their own underground water use it more efficiently than their neighbors who rely on canal water. Typically they get twice the yield, mostly because the water is there when they want it and can be put where they want it.

But how long will there be water underground? I went to see Tushaar Shah, the director of the International Water Management Institute groundwater research station, based in the Indian state of Gujarat. He has spent more than a decade following India's groundwater revolution. And at the moment, he says, Indian farmers are living in a fool's paradise. They are draining their water reserves with reckless abandon, growing thirsty crops like rice, sugarcane, alfalfa, and cotton with no thought for the future. "It's a colossal anarchy," he explains. The farmers are certainly destroying their children's future, if not their own.

The business is a free-for-all. "Regulation is virtually impossible," says Shah. "Nobody knows where the pumps are or who owns them. There is no way anyone can control what happens to them." There are no reliable statistics on how much water the farmers pump from beneath the ground, but recent

[handwritten margin note: waterless future → old end the green revolution]

estimates put the annual abstraction for irrigation at about 200 million acre-feet of water a year. That is about 80 million acre-feet more than the rains replace. This underground water irrigates at least two thirds of India's crops. It feeds India—while it lasts. But as every year passes, the underground reserves become emptier. The pumps have to work ever harder. The boreholes have to be replaced or sunk to greater depths.

"This has all just exploded in the past decade, since the arrival in India of cheap pumps. And the juggernaut is still accelerating. There are a million more pumps every year," says Shah. "We are only just beginning to see the consequences. But I must say it looks like a one-way trip to disaster." Shah estimates that at least a quarter of Indian farmers are mining underground water that nature will not replace. That is 200 million people facing a waterless future. The groundwater boom is turning to bust, and for some the green revolution is over.

Fifty years ago in northern Gujarat, circling yoked bullocks could lift water in leather buckets from open wells dug to about 30 feet. Now tube wells are sunk to 1300 feet but still run dry. Half the traditional hand-dug wells and millions of tube wells have dried up across western India. Two thirds of Tamil Nadu's hand-dug wells have failed already, and only half as much of the state's land is irrigated as a decade ago. There are 15,000 abandoned wells around Coimbatore, the state capital. Whole districts in arid states like Tamil Nadu and Gujarat are emptying of people. There has been a spate of suicides among farmers. Many more farmers are joining the millions crowding into urban slums or the gangs of construction workers and laborers traveling the roads of India.

As the water tables fall, some states find that half their heavily subsidized electricity is being used by farmers to pump water to the surface. Shah estimates than the total energy subsidy to Indian farmers to pump up underground water amounts to some $5 billion a year, more than 1 percent of gross domestic product (GDP). There is no easy way out. "If the electricity sold to farmers for pumping were charged at its full market price in northern Gujarat, nothing would grow except what the rains can sustain," he says. But the grids cannot take the strain, and there are widespread blackouts—the only effective limit on pumping, as Shah ruefully points out.

Why have Gujarat and so many other parts of India reached such a dangerous hydrological state? This is not a case of bad people doing bad things. Far from it, as I discovered when I visited Jitbhai Chowdhury, who farms on the edge of a village called Kushkal in northern Gujarat. He is, by conventional measures, a model farmer. That's why Shah's researchers took me to see him: they judged him to be the most efficient farmer for miles around. He is ecologically minded, too. He uses manure and natural pesticides made on his farm by soaking roadside weeds in water. He grows fruit trees round the edge of his fields and tends his cattle with care. I met him early one morning as the sun burned off the mist over his fields. He was milking his cows, emptying the contents of their udders into a milk churn, which he takes twice daily to a village collection point, from which trucks transport the milk to the state dairy. It seemed a perfect organic dairy farm.

But probe a little deeper and Chowdhury's very efficiency suggests the madness of the water economics being played out here. Milking done, Chowdhury described for me how his farm works. He has just 5 acres of land—land that would be virtual desert without underground water. He has a small pump that brings to the surface 3200 gallons of water an hour. It takes him sixty-four hours to irrigate his fields—a task that he carries out twenty-four times a year, mostly to grow alfalfa to feed his cows. His farm's main output is 6.5 gallons of milk a day.

I did the math. He uses 4.8 million gallons of water a year to grow the fodder to produce just over 2400 gallons of milk. That's 2000 gallons of water for every gallon of milk. According to Shah, that is better than the local average, which varies between 3000 and 4600 gallons. Even so, calculated over the year, it means he pumps from under his fields twice as much water as falls on the land in rain.

No wonder the water table in the village is 500 feet down and falling by about 20 feet a year. What looks at first sight like an extremely efficient local economy, making milk in the desert for a dairy that trades across India, is in fact hydrological suicide. Some have called the dairy industry here a "white revolution." But, says Shah, "The dairy industry is one of the major reasons for the water crisis in Gujarat." He calculates that two districts alone are export-

38

" if I don't pump, my neighbor will"

A difficulty in acting for the future - as opposed to reaching to present concerns

why so diff. to break this cycle

ing from the state 1.2 million acre-feet of virtual water a year in the form of milk.

And Chowdhury, the dairy farmer, understands all too well the bind that he and his fellow water plunderers find themselves in. "Yes, I'm worried that the water will disappear," he told me. "But what can I do? I have to live, and if I don't pump it up, my neighbors will." As we gave him a lift into the village to deliver his milk churn, he added, "I don't want my son to do farming. I want him to get a job in the city." No wonder.

Suresh Ponnusami sat back on his large wooden veranda by the road south of the Indian textile town of Tirupur. He was not rich, but for the owner of a 2.5-acre farm he was doing well. He had a phone and a television. His traditional white Indian robes had been freshly laundered that morning. As we chatted, I notice an improbably large water reservoir at the side of his house. And then a tanker drew up on the road. The driver hauled a large pipe over the hedge and dropped it into the reservoir. I expected the tanker to start filling the reservoir with water. This was a drought region, after all. But instead the pump was working in the other direction—the contents of the reservoir began to empty into the tanker.

Ponnusami explained: "I no longer grow crops, I farm water." He has sunk boreholes deep into the rocks beneath his fields and brings water to the surface twenty-four hours a day. He used to grow rice, but apart from cultivating a little fodder for his beloved cattle, he doesn't bother anymore. He doesn't need to. "The tankers come about ten times a day," he said. "I don't have to do anything except keep my reservoir full."

Actually, he doesn't so much farm water as mine it—from ever deeper below ground. The mine will probably soon be exhausted. The water table was 1000 feet underground and falling fast, said Ponnusami as his wife poured us a cold lime drink. "Before the trade in water got going here a decade ago, you could tap water at about 500 feet, so it's going down by about 50 feet a year." How much water is left? He didn't know, but he figured he had a way to go yet. In Mandaba, the village down the road, they had drilled to 1500 feet. "But they are starting to run out of water there," he said.

When I visited Ponnusami in late 2003, around five hundred water tankers

WATER MINING!

water usage for organic cotton?

↓ *what makes it organic?*

drove each day to this small area in the southern Indian state of Tamil Nadu. Day and night they came, adding their roar to the constant whine of the farmers' Japanese pumps. Ponnusami showed me the books. He sold his water for 200 rupees, about four dollars, for a single tankerload. Of that, he spent about 50 rupees on the highly subsidized electricity available here to farmers for pumping water. There were also bills for deepening his boreholes each time they ran dry. "But it's a good living, and it's risk-free," he said. No risk of failed crops. Well, at least not until the water runs out—for this crop will one day fail for good.

I asked where all the water went. Who around here wanted five hundred tankers' worth of water every day? He named two companies with whom he had contracts. They both manufacture textile dyes in the nearby town of Tirupur. This area is part of a region known as "the Manchester of India" because of the huge cotton-growing and knitwear industry set up by local entrepreneurs in the 1950s. They don't grow so much cotton now, because they can't compete with cheap imports from China and Bangladesh. But the processing business is still booming. Town billboards advertise industrial sewing machines and computerized embroidery, peroxide bleaching, dyeing, and the spinning of local silks.

All these factories need water—the hundreds of backyard dyeing and bleaching factories most of all. They once got their water from a giant reservoir 60 miles away on Tamil Nadu's biggest river, the Cauvery. But not much water flows down that river now. Like many Indian rivers these days, the Cauvery has been reduced to a trickle, and Tamil Nadu's main reservoir is nearly empty most of the year. As a consequence, the factories have taken to buying up underground water from local farmers. It is a trade that is growing all over India. But as we shall see, there is a twist in the tail here among the sari mills.

Around the corner from Ponnusami's farm, I came across a drilling rig by the road. Ponnusami's neighbor was probing deeper and deeper into the earth to keep the water flowing for the tankers. The neighbor had a nice two-story house that clearly hadn't been paid for from anything growing in the stony land around it. And sure enough, soon another tanker driver was loading up from a big concrete-lined reservoir, which was itself being filled from two boreholes. The driver, who collected water from several local farms, said he sold the water in town for 400 rupees—a clear 100 percent profit.

(TRAGEDY)
LAW OF COMMONS

Chase ST $$$
at the cost of destroying
LT collective future
=
NO ONE can afford to miss out on the boom

He sold mostly to the textile factories. But sometimes he toured the streets, filling householders' pots for a rupee each. Public water supplies to houses are as fitful as those to the mills, and many people rely on water from tankers to drink, cook their rice, and wash. As we talked, another tanker drove up the track to the next farm. That was three tankers pumping from three farms within 300 feet of each other on one morning in the middle of the Indian countryside.

The farmer's wife came out to survey the scene. The water trade here was probably on its last legs, she agreed. "Every day the water is reducing. We drilled two new boreholes a few weeks ago and one has already failed. But we will keep drilling till we only have enough water for drinking," she said. Surely this is madness, I suggested. Why not get back to farming before the wells finally run dry? "If everybody did that, it would be well and good," she agreed. "But they don't. We are all trying to make as much money as we can before the water runs out. If we stopped pumping just on this farm, it wouldn't affect the outcome."

It is a classic case of what environmentalists call "the tragedy of the commons." Everybody chases short-term wealth even at the cost of destroying their long-term collective future. Nobody can afford to miss out on the boom, because they will all share in the eventual bust. Some think it is what we are doing to the planet. It is certainly what is happening to India's underground water.

But the reckless plunder of the water around Tirupur is only the start of this particular tragedy of the commons. I had taken these byways at the suggestion of a local academic, the director of state university's Water Technology Center, Kuppannan Palanisami. "For the real tragedy," he told me, "you have to follow what happens to the water that these farmers sell to the factories." So I followed. The two hundred or so dyeing and bleaching factories that take the farmers' water stretch almost 20 miles from Tirupur along the banks of the Noyal River, a tributary of the Cauvery. The Noyal should be completely dry for most of the year, because, owing to all the pumping, there are no springs to sustain its flow through the eight-month dry season. But it was not entirely dry. It ran with the stinking, brightly colored effluent pouring from the facto-

41

ries. This is what had become of the farmers' water after it had served its purpose in the textile works.

I did what Palanisami said and followed the foul, untreated effluent down the river. It eventually ended up in a big water-supply reservoir that had been built by the Tamil Nadu authorities a decade before, at a time when there was still clean water in the Noyal. The reservoir is in the shadow of a famous temple on a hilltop at Avanashi. It was dug to help farmers in a wide area around Avanashi grow crops all year round. I climbed onto the bank surrounding the reservoir. But one whiff, one look at the chemical encrustation on its shoreline, made it clear that its contents were poisonous. Almost since it opened, I later learned, the reservoir had been little more than an industrial sump in the middle of the countryside.

And a very leaky sump at that. The reservoir has no lining, and year by year the poison in its waters has been seeping into the local soil and down into the underground water reserves. So far, said farmers, the poison has spread for a mile on either side of the reservoir, and for about 6 miles downstream along the river. All around the dam, I found derelict houses. The fields were barren and empty. No crops would grow, or even weeds. Only old plantations of coconut trees kept going, and they looked sick and stunted.

The only functioning village I could find in this wasteland survived by processing the coconuts to make oil. "Before the dam we had good, profitable crops, like turmeric and banana. We thought the dam would help us grow more, but it turned out to be the opposite," said the village headwoman, Kittusami Manonmani. "Now we only have coconuts, and even then new trees don't grow."

Surrounded by coconut shells, she warmed to her theme. The reservoir had polluted the underground water for miles around, she said. "Our wells are full of salt now. It is not natural salt; it comes from the chemicals in the reservoir. You can't even bathe in it, let alone drink it. If we wash our clothes in it, they get holes and we come out in rashes and start itching. At the start we let the cattle drink the water, but they died." She brought me a pot of well water that had been sitting in the yard all day so the salts would settle out. There was a thick sludge at the bottom, but the water still tasted bitter. We spat it out quickly.

The local hospital in Tirupur later confirmed that there are high rates of skin disease and lung disorders in the area as a result of the chemical waste. The villagers said they had made numerous protests about their water. "Politicians used to come here and make promises to stop this. But they don't even make promises now. We think the textile companies have got to them," said Manonmani.

I can't confirm that, but during the 1990s there were inconclusive legal actions against the Tirupur factories, which at one point promised to build treatment plants. But the effluent has kept coming. In 2002 the controversy revived when local press reports caused a flurry of concern at the State Pollution Control Board. Officials revealed that 21 million gallons of dyeing and bleaching effluents entered the Noyal River each day. Sodium in the effluent was making the local soils infertile. Scientists took some water samples from the villagers. They found peroxide, hypochlorite, and benzidine, a cancer-causing chemical used in manufacturing many dyes. But nobody told the villagers the results.

When I visited, eighteen months after the reports, nothing had changed. With no sign of a cleanup, more and more villagers were leaving the area. Even in the coconut village, more than half of the men had left to get jobs elsewhere. Inevitably, perhaps, that meant knocking on the doors of the textile factories of Tirupur, the cause of their misery.

How do the villagers left behind live? I asked Manonmani where they got clean water to drink and wash. "We have to buy the water from tankers," she told me. "They come from that way." She pointed west and south. "About 10 miles, I think." And then it dawned on me: the tanker drivers who sold water to these villagers were the same tanker drivers who bought the water from the water-mining farmers on the other side of town.

These villagers in this toxic wilderness were buying their water, at the price of a rupee a pot, from the people who sucked dry the precious underground reserves of Mandaba. The same people who were keeping the dye factories in business, producing the effluent that poisoned their fields and wells for miles around, were making a further tidy profit out of the misery caused by their pollution. The tragedy of Tamil Nadu's disappearing water supplies was complete.

———

Across India there are similar bizarre local dramas played out over water short-ages. Water disputes in one village in Kerala, in the far southwest of India, be-came an international news story in 2003, when angry farmers accused the local Coca-Cola factory of drying up their fields to fill its bottles. The village of Plachimara took the company's biggest bottling plant in Asia to the state high court on a charge of destroying their coconut groves and paddy fields. Since it opened in 1998, the plant had taken around 130,000 gallons a day from a series of boreholes on its 40-acre site.

The story got taken up by the international media. You can see why. The amount pumped sounds like a lot, and a global corporation like Coca-Cola is an easy target. But the truth is that even 130,000 gallons of water a day is not large on the scale of rural Indian water use. A single local rice farmer with 25 acres of paddy could easily be using as much water as the Coca-Cola bottling plant. The villagers who complained that Coca-Cola was emptying their wells were almost certainly collectively using more water to irrigate their crops than the company was taking to fill its bottles.

In late 2003, the Kerala court told Coca-Cola to stop pumping out the lo-cal water and find an alternative source. The plant subsequently shut. The de-cision was later reversed, but the village council claims that in law it has the final say, and the case went to the Indian Supreme Court. The plant remained closed in late 2005. Beyond the legal issues, the key hydrological question is, will the water come back? If, as seems likely, it does not, then it will underline a frightening truth: that the state of Plachimara's wells is not a gross example of a global corporation riding roughshod over locals but simply a typical story of what India is doing itself to its most precious resource.

6

Halliburton's Job for Qaddafi

Some say ancient underground water that is not being replaced by the rains, which is often called "fossil water," should never be pumped except in dire emergencies. It should be maintained as a precious backup reserve. Others argue that new water reserves will always be found. Just as shortages of fossil fuels and precious minerals have encouraged geologists to locate new sources, they say, so it will be with water. The problem with this argument is that water does not command the kind of prices that enable it to be transported around the world as other commodities are. Water still needs to be local. Unless you are Colonel Qaddafi, that is.

Libya, the driest nation of its size on earth, has a river at last. To be precise, it has two rivers, though they go by the singular name Great Manmade River. Neither is a river you can go fishing in, or watch flow by as you eat a picnic on the bank. They don't have banks, or fish. The rivers comprise a 2000-mile network of pipes, all of them wide enough to drive an underground train through. Their water comes from hundreds of boreholes sunk up to a third of a mile into the sand in the middle of the Sahara Desert.

The giant Nubian sandstone aquifer is the largest liquid freshwater source on earth, containing around 50 billion acre-feet of water in a series of basins whose only outlet until recently was a handful of oases in the desert. Some of the water, according to recent radiocarbon dating, is more than a million years old. Most of it comes from a more recent wet era when the Sahara was crocodile-infested swamp. But that era ended abruptly seven thousand years ago, when the desert formed. And since then, except for a little infiltration from the Tibesti Mountains in northern Chad, it has sat undisturbed for thousands of years. Until now.

Geologists have eyed this water ever since discovering it in the 1960s, while prospecting for oil. Libya's leader, Colonel Muammar al-Qaddafi, who seized power in 1969, decided to spend his oil revenues to bring the water to the surface. At first he tried to persuade Libyans to move to his desert boreholes and grow wheat. But they wouldn't go. They preferred their fields along the Mediterranean coast, even though the aquifers that watered them were emptying fast. The remains of Qaddafi's giant desert farms can still be seen from aircraft crossing the desert.

So the eccentric colonel decided instead to send the water to the people. The first phase of the Great Manmade River opened in 1991, carrying 600,000 acre-feet of fossil water each year from well fields around the oases of Sarir and Tazirbu on a nine-day journey across 600 miles of desert so dry that even camel trains didn't cross it until the nineteenth century.

There was a huge ceremony on the coast near Sirte to greet the first water. Fellow leaders from Africa and the Middle East gathered in the cool desert night to hear the colonel compare his great work to the pyramids of ancient Egypt and watch him unleash the water into a reservoir. It had cost Qaddafi $14 billion to deliver that water. No wonder a thousand Libyans rushed forward to take ecstatic sips. Qaddafi boasted that day that his new river was a revolutionary triumph against the tyrants of the United States and Britain, who had bombed Tripoli just five years before. He forgot to mention that the project was being masterminded by the enemy.

The project's headquarters was an office on the fourth floor of a top-security high-rise near the train station at Hampton Wick, in the leafy suburbs of southwestern London. There, through the bombings and kidnappings and sieges and murders that characterized Anglo-Libyan relations in the 1980s, and a U.S. embargo on trade with Libya, a European subsidiary of the American engineering corporation Halliburton called Brown & Root was legitimately able to work with Qaddafi's staff to make the colonel's pipe dream come true. When I visited Bashir el-Saleh, the colonel's U.S.-trained head man in Hampton Wick, he said, "This is the best investment we have ever made. Hopefully we will be able to depend on it for hundreds of years."

Since then, a second phase of the "river" has been completed, along with a series of coastal connections. The second phase is of a similar size to the first but had to cross mountains and canyons to link the capital, Tripoli, with the western well fields at Fezzan. It proved more difficult and has reportedly been delivering only a fifth of its intended flow since going onstream in the late 1990s.

So far, half a million concrete sections, each 13 feet wide, 23 feet long, and painted Islamic green, have been assembled. The project, for some years the world's largest civil engineering project, has used some 5 million tons of cement (enough to pave a road from Tripoli to Bombay, say the engineers proudly) and 25 million tons of aggregate. Three more phases are planned; they will link up the two completed rivers and extend the first out to Kufra.

But progress has slowed since the other key collaborator, a giant South Korean steel company called Dong Ah, responsible for making the pipes, went out of business in the aftermath of the Asian financial crisis of the late 1990s. And several times corrosion has caused the partially buried pipes to burst, flinging pieces of pipe a hundred yards across the desert, killing any passing camels, and forcing the closure of the "river" for months of repairs.

So far, Qaddafi has spent about $27 billion on the project. It has drained Libyan oil revenues for more than two decades. So is this a fantasy born of oil money and the megalomaniacal dreams of Colonel Qaddafi? Or could this be the future? Could the twenty-first century see the construction of many more artificial rivers to tap new reserves of water and take it where the people are?

The project is certainly shot through with fantasy on all sides. The Libyan authorities see it as an "eighth wonder of the world" and a source of "everlasting prosperity" (a term first coined by another colonel, Egypt's Gamal Abdel Nasser, to describe his own engineering wonder of the world, the Aswan High Dam). Meanwhile, conspiracy-minded foreigners have developed their own bizarre notions. The *New York Times* and the *Guardian* in 1997 splashed a story quoting defense analysts who claimed that the pipes were not intended to carry water at all. They had been built, the papers said, to provide camouflage for Libyan tanks in some future invasion of the country's neighbors and to provide Qaddafi with "somewhere to store chemical and biological weapons."

But Qaddafi does have a serious purpose in all this. The aim is to deliver 1.6 million acre-feet of water a year to help Libya feed its people. The country has more than a million acres of irrigated fields. But most are on the coast, where overpumping and an invasion of seawater into aquifers has left irrigated soils so salty that many wheat fields have perished. Oranges are said to grow no bigger than Ping-Pong balls. Partly as a result, Libya currently imports half its grain. But unlike neighbors such as Egypt, it persists in a goal of agricultural self-sufficiency. No virtual water for Colonel Qaddafi.

The trouble is, the numbers simply do not add up. The embarrassing truth is that although $27 billion has been spent on taking desert water to the coast, most of Libya's irrigation water still comes from the saline coastal aquifers. Irrigating the existing farmland at current standards of efficiency requires around 4 million acre-feet, more than twice the intended output of the man-made rivers. The flow down Qaddafi's pipes is actually rather puny. For all the talk of a "New Nile," phase one delivered about as much water as England's River Mersey, and all five phases together will deliver roughly the flow of the Thames through London. "We would need two or three times more to be self-sufficient in food," says Abou Fayed, who is responsible for allocating the water to farms.

Qaddafi's reach seems to have exceeded his grasp. The vast capital cost and the growing bills for pumping water from ever greater depths beneath the desert make wheat grown with the Saharan water some of the most expensive on earth. "It is madness to use this water for agriculture," says Tony Allan, a British specialist on Middle East water. In future, as Libya begins the painful process of integrating itself into the global economy, pressure is likely to grow for it to import more wheat rather than less, and to reduce rather than increase its irrigation. It seems unlikely now that the Great Manmade River will ever be fully utilized. In the end, it may be left to rust in the desert.

7

The World's Largest Mass Poisoning

I met ten-year-old Shatap on a muddy lane in Hirapur, a village in central India. He had a gait straight out of Monty Python's silly walks (I hope that is not an offensive remark, but it is how it struck me at the time). The diminutive figure was not playing games as he waddled up the lane, his knees locked together and his stunted and misshapen lower legs splayed wide, almost like flippers. His disability was permanent. He had been poisoned by the water in the well at the village school. His bones were grossly deformed by fluoride dissolved in the water.

The school tube well at Hirapur was one of tens of millions sunk around the world in a highly publicized race to provide the world's poor with "safe" drinking water. It was planned and partly funded by the UN children's fund, UNICEF, during the International Water Decade, the 1980s. The decade worked on a simple premise. Millions of the rural poor drank water taken untreated from surface sources such as rivers, ponds, and canals. This water was polluted, often with sewage that spread lethal diseases such as cholera and typhoid and that caused diarrhea. But while surface water was often polluted, underground water, because it had not come into contact with manmade pollution, was almost universally safe. So sure of this were the engineers and public health professionals behind the campaign that they often didn't even run basic tests for poison in the water as the wells were sunk.

Such folly. They forgot that nature can poison water too. And now the truth is dawning. Shatap is one of perhaps millions of drinkers from the wells sunk by the UN and other aid agencies who are suffering the symptoms of fluoride poisoning. Next door to Shatap, I met fourteen-year-old Krishna Chudaman and her sister, Kamala. Both had similar symptoms. Bowlegged Krishna had given up education because, after being poisoned at her primary school, she couldn't walk to the secondary school in the next village. In fact, almost every child in the village had some symptoms. Hirapur is one village among dozens to suffer in Mandla District, a small corner of Madhya Pradesh.

In nearby Tilaipani, the cattle went lame before an epidemic of knock-knees among children alerted a young researcher to the problem, which local doctors had misdiagnosed as rickets. UNICEF funded the wells, said Ravi Shankar Tiwary, the director of the local medical center. Government engineers claimed to have tested the water, but "I know they didn't—they didn't have the right equipment," he said. He showed me a file full of X-rays of the limbs of young children, bent like bows. "As they grow, it becomes worse and they are crippled," he explained.

Fluoride is a common component of the granite rocks that underlie much of India. It slowly dissolves into the underground water in the pores of rocks above it. Usually the fluoride doesn't spread far from the granite, so wells that tap water near the surface are generally safe enough. But as water tables fall, villagers sink their wells ever deeper and ever closer to the bedrock. As they do, the chances of bringing up toxic water dramatically increase.

A lot of misery gets overlooked in rural India, a country of a million villages. Nobody knows how many suffer from fluoride poisoning. But Andezhath Susheela, of the Indian government's Fluorosis and Rural Development Foundation in Delhi, guesses that 60 million could suffer to some degree. Unless public health officials show much greater vigilance in testing well water and finding alternative supplies when fluoride is discovered, the crisis is bound to worsen as the groundwater crisis itself deteriorates, she told me.

"It's a huge time bomb," agreed Tushaar Shah, at the International Water Management Institute in Gujarat, a state badly hit. "And in many areas there is really no solution. Going back to polluted surface water would be even worse. We are looking into developing a fluoride filter, but the impracticalities of using these in small villages where they can't get the chemicals are huge."

From Assam in the east to Gujarat in the west, from Tamil Nadu in the south to Kashmir in the north, doctors are starting to blame outbreaks of anemia, stiff joints, kidney failure, muscle degeneration, and cancer on fluoride in the water. In two districts in Rajasthan, more than half the population has symptoms. The greatest risks seem to be among the young and the old. The young suffer damage to their limbs as they grow; the old have the highest accumulations in their bodies. In western Gujarat, I met elderly hunchbacks and cripples whose problems were almost certainly due to fluoride.

And India is not alone. China is estimated to have more than a million victims. From Chile to Ethiopia to Uzbekistan, similar stories are probably waiting to be told in tiny villages where simple hand-pumped tube wells have been installed to provide "safe" drinking water.

As amazing as it may seem, there is an even bigger scandal that has been created by the trend of drinking water from underground sources: arsenic. Tens of millions of people across Bangladesh and western India are drinking well water laced with concentrations of arsenic that are likely to kill them eventually. The World Health Organization (WHO) calls it "the largest mass poisoning of a population in history." At least half of Bangladesh's estimated 12 million backyard tube wells are thought to be poisoned. Many deliver water with hundreds of times the WHO limit for arsenic in water supplies.

The arsenic began in the rocks of the Himalayan Mountains but was eroded by great rivers like the Ganges and the Brahmaputra and washed downstream, probably over thousands or millions of years. The toxic metal gradually accumulated in the thick muds of the floodplains and deltas that make up most of Bangladesh. And there it stayed undisturbed, until humans in the past thirty years began to pump up water out of the mud for drinking. The water brought the arsenic with it. Tube wells sunk to depths of between 60 and 300 feet contain the most. Catastrophically, that is just the level to which most tube wells have been sunk.

This is a silent epidemic that is stalking the majority of Bangladesh's 68,000 villages. Arsenic, like fluoride, is a cumulative poison. It typically takes a decade of drinking arsenic-laced water for the first symptoms to appear. And nobody knows who is at risk. Virtually every backyard has a tube well, and seemingly small differences in location and depth can make big differences in the arsenic content of the water. Almost any tube well could be delivering lethal concentrations. To be sure, each well has to be tested. Very few have been. Tens of thousands of Bangladeshis have already developed skin lesions, cancers, and other symptoms. Many have died, and a quarter million could succumb in the coming decade, says Allan Smith, an American doctor who first alerted the WHO to the scale of the scourge.

Tube wells were installed in Bangladesh and elsewhere to reduce the heavy death toll from sewage-borne bacteria and diseases spread by contaminated surface waters. In the 1970s, these diseases were killing an estimated 250,000 people a year in Bangladesh. Many people are alive today as a result of the sinking of the tube wells. But the arsenic epidemic could have been prevented if the risks had been realized in time. Tube wells could have been sunk deeper, to a depth where the water is generally free of the poison, or the water could have been tested before being used for drinking.

The true scale of the disaster only emerged slowly during the 1990s. But tragically, ever since then, efforts to undo the damage have been foiled by indolent local bureaucrats. It emerged in 2003 that efforts to find and replace the estimated 6 million poisoned tube wells were making little headway. The Bangladeshi government had spent less than $7 million of the $32 million given by the World Bank in 1998 for an immediate cleanup. But it is too easy to blame Bangladeshi bureaucrats. The truth is that thousands of people are still being poisoned by tube wells sunk with aid money from the British and other governments, from charities, and from UN agencies like UNICEF, which sunk the first 900,000 wells.

———

I went out on a survey expedition to find poisoned wells. We chose our first village at random. Like thousands of others across Bangladesh, Dipordi, a couple of minutes out of Sonargaon, the ancient capital of Bengal, had never had its water tested for arsenic, and none of its inhabitants had been screened for the debilitating skin diseases that are its first symptoms. Like a traveling medicine man, Akhtar Ahmad, of the National Institute of Preventive Social Medicine, walked unannounced into the village. He called for a tumbler of water from each well and set out his test kit on a hastily erected trestle table.

Within half an hour, before an enthusiastic crowd, he had analyzed samples from nine wells for arsenic, using test tubes, four chemical reagents, a clamp, and test papers that darkened from yellow to brown as the arsenic level rose. It was part circus and part science. But it revealed seven poisoned wells and only two that were safe. One was off the scale, with more than 500 parts per billion (ppb) of arsenic, fifty times the recommended WHO limit and ten times the Bangladesh government's own limit.

The results announced, Ahmad lined up the crowd to check their palms, chests, and gums for signs of the sores typical of arsenic poisoning. These fifteen-second consultations yielded five cases—a small hit rate by local standards, but Ahmad was worried. Villagers said the first tube well had been installed here only seven years before. Symptoms of arsenic poisoning typically take ten years to show up. Yet, Ahmad said, "Already we find symptoms. This is the first phase of something big." One villager was angry that international aid had proved such a poisoned chalice and asked about installing a deeper well to bypass the arsenic-bearing water. "I am rich; I can afford it," he said.

But it was too late for Abdul Kasem, whom we met a few miles away in Barai Kandi village. Kasem had turned up at a clinic a week before with a gruesome carcinoma on one hand and a flask of well water in the other, asking for checks on both. His medical diagnosis had been instant, but we had returned to give Kasem the test results on his water. Sure enough, the well from which he had drunk for twenty years contained a lethal 500 parts per billion of arsenic.

Kasem, his carcinoma covered by a leaf from the field, took us to meet his wife and five children. All six, along with another twelve villagers living in nearby houses, had warts and other growths characteristic of arsenic poisoning. Ahmad gave them prescriptions for vitamins A, C, and E. "It might help, and it's all I can do," he said. Meanwhile, these villagers assembled water samples from thirteen wells and joined in the testing: 600 ppb, 300 ppb, 30 ppb, 1000 ppb...Only two samples passed the Bangladeshi limit, and none the WHO limit. Before we left, an impromptu village meeting had forced the owner of the two cleanest wells to let his neighbors take drinking water from him.

Ahmad's work was the first evidence I had seen, after nearly a week in the country, that there were people in the Bangladeshi government's service who possessed the will and skills to tackle an epidemic of arsenic pouring from millions of wells in thousands of villages like Barai Kandi and Dipordi. "We have to act ourselves to find the safe wells in each village," Ahmad said. "We don't need international loans to do this, just commitment." His test kit was cheap and crude and far from accurate by laboratory standards, he admitted. "But we can test a sample for seven thaka [about ten cents] and give the results instantly to the village," he said.

This is barefoot science. You couldn't publish scientific papers using such "dirty data." But it is probably the only practical way that villagers will be able to take back some control of their health from the public health professionals who have so damaged it. Ahmad told me he wanted the kit mass-produced so every village could have one. "It is very simple to use. Everything is color-coded, so people don't even need to read," he said. "Our work is about empowerment. We want the villagers to be able to say, This is our problem, we will solve it." The difficulty in making that happen was that the kit was so cheap and simple that no company wanted to manufacture it. There were no profits to be made.

The arsenic epidemic keeps growing. High levels of arsenic in groundwater were once believed to be confined to an area around the Ganges delta in Bangladesh and West Bengal. But new studies in 2003 suggested that arsenic might be lurking in underground waters near the river all the way to the Himalayas—across a region where 800 million people drink underground water.

The new focus of concern was the state of Bihar, home to 83 million people. There, Kuneshwar Ojha, a schoolteacher living in a tiny village called Semria Ojha Patti, close to the Ganges, had become concerned after his wife and mother had both died of liver cancer. He noticed that other family members had developed skin lesions, and he thought they looked like symptoms of arsenic poisoning. He took water samples from the family tube well to Dipankar Chakraborti, the director of environmental studies at Jadavpur University in Calcutta, who had originally uncovered the mass arsenic poisoning in West Bengal.

Chakraborti confirmed high concentrations of arsenic in the sample. And it quickly emerged that eighteen young people had died from apparently arsenic-related illnesses in the village in the previous five years, and a hundred more were sick with early symptoms. The only fit people were the Dalits, or untouchables, who were not allowed to drink water from village tube wells because of their low status.

Since then, hundreds of similar cases have been found in the surrounding villages in this part of Bihar, and the authorities have banned people from using many tube wells. Parts of Semria Ojha Patti were abandoned because there were no safe wells. Chakraborti surveyed wells across a wide area of the state and found that 40 percent contained arsenic above the government limit—a proportion very similar to that in Bangladesh. More than half the adults examined in the study showed symptoms of arsenic poisoning.

A few months later, similar findings began to emerge from neighboring Uttar Pradesh. "The pattern we saw in Bangladesh is being repeated," Chakraborti said. "There we began with the discovery of it in three villages. Now thousands are known to be affected, and more are being discovered all the time. Our early warnings were ignored then. Now we are warning about Bihar. We feel that this is just the tip of the iceberg."

Is there anything special about the Ganges? Maybe not. Within weeks of the public reporting of Chakraborti's new findings, Michael Berg, of the Swiss Federal Institute for Environmental Science, reported on a study of water from tube wells sunk beneath the Red River delta in Vietnam. They contained arsenic levels up to three hundred times the WHO safe limit. Symptoms of arsenic poisoning could soon appear, he said. The first tube wells had been installed seven years before. The delta is home to 11 million people, including most of the inhabitants of the capital, Hanoi.

8

Mirages

Over the past twenty years, tens of millions of small farmers across much of the poor world have been pumping water from beneath their fields. As in India, it is an extensive revolution born of two factors: first, the failure of government-built irrigation systems that tap rivers to deliver the water the farmers need; and second, advances in technology that allow them to drill far deeper into the earth for water than they could with their old hand-dug wells and to buy cheap Japanese pumps to bring the water to the surface.

There are few official statistics, and there is no way of collecting them. But just three countries—India, China, and Pakistan—probably pump out around 325 million acre-feet of underground water a year from the new tube wells. They account for more than half of the world's total use of underground water for agriculture. And they are living on borrowed time, sucking dry the continent's water reserves.

Every year perhaps 100 million Chinese eat food grown with underground water that the rains are not replacing. There are, as we have seen, another 200 million or so doing the same in India. In the Pakistani province of Punjab, which produces 90 percent of that country's wheat, farmers compensate for diminishing deliveries of water from the Indus River by pumping from beneath their own fields. They pump 30 percent more than is recharged, and water tables are plunging by three to six feet a year.

Overall total pumping in India, China, and Pakistan probably exceeds recharge by 120 to 160 million acre-feet a year. The boom has so far lasted twenty years; the bust could be less than twenty years away. The consequences of the eventual, inevitable failure of underground water in these countries could be catastrophic. But it is a crisis that has not yet registered on the radar screens of governments or aid agencies. When a river runs dry, it is very visible. But underground water is invisible. Only the farmers know they have to drill deeper and deeper to find it. And few in the corridors of power talk to farmers about a slow-burning disaster that will one day affect hundreds of millions of people.

It won't happen everywhere at the same time, of course. Each aquifer has its own countdown to extinction. But even so, as Tushaar Shah puts it, "The overuse of water in Asia's underground aquifers will spell disaster for millions of the region's poor people, who depend on it." And as each aquifer dries up, it will undermine the world's ability to feed itself.

Farmers in other heavily populated countries are beginning to sink tube wells and buy pumps with equal enthusiasm. In the past decade, Vietnamese farmers have quadrupled the number of tube wells to more than a million. Sri Lanka, Indonesia, Iran, and Bangladesh are not far behind. These countries are at the heart of what the agronomist and environmentalist Lester Brown calls "Asia's food bubble." Record farm outputs in recent years, he says, have been made possible only by an unsustainable assault on this fast-diminishing resource. It is a bubble that is bound to burst. "The question is not if, but when," he says.

Asia leads the way in this unsustainable groundwater revolution, but countries such as Mexico, Argentina, Brazil, Saudi Arabia, and Morocco are increasingly significant players. Sub-Saharan Africa is not far behind. As rivers fail, underground sources provide a third of the world's water. By some calculations, as much as a tenth of the world's food is being grown using underground water that is not being replaced by the rains. Major cities—among them Beijing and Tianjin, Mexico City and Bangkok—are also growing increasingly reliant on pumping out underground reserves.

As water tables sink, the revolution is putting a premium on finding new underground water reserves—preferably ones that are recharged by the rain. New water is still being found in the bowels of the earth. Chinese scientists recently reported the discovery of unsuspected water beneath a vast system of sand dunes in the Gobi Desert in Inner Mongolia, which they claim is being replenished from mountains to the south. And hydrologists reported new research into the likely size of the Guarani aquifer, which stretches beneath more than 400,000 square miles of Brazil, Argentina, Uruguay, and Paraguay. It may contain 40 billion acre-feet of water—as much as flows down the Amazon River in seven years. And it is still being replenished by the rain. The aquifer is already supplying 15 million people with water without any general decline in water tables. Hydrologists now believe it could one day supply as many as 200 million people. They want to build an aqueduct to take the water to the world's third largest city, São Paulo.

———

But as fast as one door opens, another closes. The High Plains of the Midwest are part of American history. The first white settlers going west transformed the plains from buffalo hunting grounds into rough pasture for cattle. It was the land of the cowboy. Then came the plow, and pasture became dry prairie— until the drought of the 1930s blew the soil away. Since those Dust Bowl days, when millions of sharecroppers abandoned the land and trekked on to California, the arid plains have been transformed once again through the pumping of water from a giant underground reserve discovered beneath them.

Known as the Ogallala aquifer after the Sioux nation that once hunted buffalo here, the reserve stretches beneath most of Nebraska, southern South Dakota, western Kansas, Oklahoma, and Texas and parts of eastern New Mexico, Colorado, and Wyoming. In the 1930s, as the Dust Bowl refugees poured into California, just 600 wells tapped this aquifer. But by the late 1970s there were 200,000 wells, supplying 22 million acre-feet a year to more than a third of the country's irrigated fields.

The aquifer was an enormous U.S. resource, but also a global one. In a good year, the High Plains produced three quarters of the wheat traded on the world market—restocking empty Russian grain stores, feeding starving Ethiopians, and keeping Egyptians fed as the Nile ran dry. The United States became the world's biggest exporter of virtual water—but at the expense of draining the Ogallala.

The problem is that much of the aquifer is to all intents and purposes a fossil resource laid down in wetter times. Little water is added to it from the region's scanty rains. And starting in a few isolated southern pockets, wells have been drying out now for more than thirty years. The first to hear the news, recorded a U.S. water researcher named Sandra Postel, were the farmers of Deaf Smith County, in the Texas panhandle. During the summer of 1970, a well that had been pumping water since 1936 suddenly went dry. And many others have followed.

Today more than a quarter of the aquifer is gone in parts of Texas, Oklahoma, and Kansas, and over wide areas the water table has fallen by more than 100 feet. All well-sinking is banned in some places, and fewer than 10 million acre-feet are now pumped annually, less than half the output in the 1970s. Fly over the land and you can see the circular marks where rotating sprinklers once kept the soil wet and the fields green—but the soil is now dry and the fields are brown. The sagebrush and buffalo grass are returning. The buffalo may follow.

Nationally, the United States is not heavily dependent on underground water. Aquifers provide less than 1 percent of water supplied in New England, for instance. But a third of all its irrigation water does come from underground, and some states in the South and West would be more or less literally lost without it. The country's three major aquifers—the Ogallala, the Central Valley aquifer in California, and the Southwest aquifer—are all in the arid West. The Southwest aquifer in particular is vital. In Arizona, it is virtually the only water source within the state. No wonder that, across the West, cities are buying up farms to get their rights to pump these aquifers.

But overpumping in the West is as widespread as on the High Plains. For many years Arizona has been removing underground water at twice the rate that rains can replace it. In California people pump out 15 percent more than the rains replenish—an overdraft of 1.3 million acre-feet a year. The combined annual overpumping of the Ogallala, Central Valley, and Southwest aquifers has been 30 million acre-feet, resulting in a cumulative total of over 800 million acre-feet in recent years. And as the aquifers empty, the honeycomb rocks that once held the water in their pores are gradually being crushed by the rocks above. One estimate is that this crushing has permanently reduced the water-storage capacity of the Central Valley aquifer by half—the equivalent of blowing up a dam the size of Grand Coulee.

———

Some of the most serious, indeed scandalous, groundwater disasters are being played out in the Middle East. Saudi Arabia has virtually no rain and no rivers or surface lakes of any kind. It has spent $10 billion on desalination works to fill its faucets. But during the 1980s, the government spent another $40 billion of its oil revenues sinking pumps into the vast aquifer beneath the desert and marking out 2.5 million acres of desert for wheat farms to use the water. All of the water was provided to farmers free of charge. Sprinkler arms 300 yards long swept over the sands, distributing water pumped up from as much as half a mile below. Nobody cared how much water was wasted, and usually most of it evaporated in the sun. For every ton of wheat grown, the government supplied 2.5 acre-feet of water—three times the global norm. It was an operation that, as the *Economist* suggested at the time, "made about as much economic sense as planting bananas under glass in Alaska."

What to do for an encore? Since the mid-1990s, many Arabian farmers have converted their wheat fields to alfalfa fields—which use even more water—to fill the feedlots for a new national obsession with dairy cattle. This time, high-tech cowsheds have sprouted across the desert. Inside them, fine mists of water control temperatures in the baking desert sun. Just as wheat once grew in the desert, now Holstein cattle are milked in the land of the camel.

There is undoubtedly a lot of water beneath the Arabian peninsula, but much of it is very deep and very old. Virtually none of it will ever be replaced. At the start of the 1980s, the country was calculated to have proven underground reserves of over 400 million acre-feet. But since the mid-1990s, that water has been removed at a rate of more than 16 million acre-feet a year. Saudi experiments in farming the desert have so far wasted 240 million acre-feet of water, says Elie Elhadj, of the School of Oriental and African Studies in London. Some 60 percent of the water is gone.

Saudi Arabia is slowly waking up to its profligacy. In 2004 it launched a water conservation drive aimed at halving domestic consumption of its desalinated seawater. Crown Prince Abdullah publicly replaced his toilet tank with a more efficient model. More significantly, water vice minister Abdullah al-Hussayen promised a root-and-branch review of agricultural water use. "We need to examine whether we need to produce the whole domestic consumption of wheat or perhaps make do with half," he said.

Meanwhile, Saudi pumping is raising tensions with the country's neighbors. Jordanian hydrologists fear that the Saudis are draining the Disi aquifer, which lies beneath their joint border. Jordan is one of the driest states on earth and lacks cash for desalination. Disi represents its "last substantial unexploited water resource," according to a UN study, which said the aquifer should be kept "for emergency use only." There's no chance of that. While the Saudis pump as hard as they can from one side of the border, Jordan is rushing ahead with plans to pipe Disi water 200 miles north to the capital, Amman. Whoever gets the water, the aquifer seems doomed.

———

Whatever Jordan may face in the future, Gaza faces now. The densely packed Palestinian enclave, a narrow coastal strip between Israel and Egypt, is running out of drinkable water today. Aside from Kuwait, which lives on desalinated seawater, Gaza's million-plus inhabitants have the lowest per capita availability of natural freshwater in the world.

Palestinians pump at least 100,000 acre-feet a year from their aquifer beneath the sands. The recharge rate is about 73,000 acre-feet. Not surprisingly, the water table is plunging. Meanwhile the porous rocks are being invaded by sewage from Gaza's towns and refugee camps and by salty seawater from the coast. The aquifer's bacteria count and salinity level rise inexorably, making the water nasty to drink and increasingly poisonous for crops.

The crisis is partly of the Palestinians' own making. Two thirds of the water is brought to the surface by Palestinian pumps for irrigation, much of it from wells never licensed by the Palestinian Water Authority. But they are not the only people with pumps. Israelis farm the Negev Desert right up to the Gaza border. Seen from Gaza, their greenhouses glint in the sun, guzzling up water that could be growing Palestinian crops.

Some clean water does enter the Gaza aquifer. Some arrives as rain, and some comes down the Wadi Gaza. This is the coastal end of a desert drainage basin that begins in Palestinian territory on the West Bank south of Jerusalem, then passes through the Negev Desert in Israel and flows to the sea through Gaza, where it replenishes the aquifer through a small coastal wetland. Or it did. The Israelis have built dams on the wadi (which in the Bible and on Israeli maps is called Besor). These days, little water ever reaches Gaza. Most of it is grabbed to irrigate Israeli fields. And the wetland has turned into a sewage sump.

The international community has plans to improve Gaza's water supplies by building infrastructure to recycle irrigation drainage water and catch more of the sparse rains. The U.S. government also wants to build a desalination plant. "It may take twenty to thirty years, but we'd hope to get the situation into equilibrium, with abstraction from the aquifer no greater than recharge," Alvin Newman, the U.S. point man, told me in his high-security offices in Tel Aviv.

But there may not be enough time, says Clemens Messerschmid, a German hydrologist working for the Palestinian Water Authority and the British government's aid agency. UN studies suggest that the aquifer could be effectively dry within fifteen years. The situation "is a catastrophe. It is an emergency case," according to Messerschmid. To save the aquifer, he says, the Palestinians have to close their wells and the Israelis have to supply water. "It is as simple as that. Israel has plenty of capacity in the south of the country. The Gazans will have to pay, of course, but Israel should supply."

During the intifada, the continuing disputes and armed conflicts in Gaza got in the way of action to halt the catastrophe. Everything was "on hold," I was told. But whatever the politics, the hydrology will not wait.

III

When
the rivers
run dry...

the wet places die

9

The Common Wealth

From the peat bogs of Scotland to the backwaters of the Mississippi delta, from the lagoons of Venice to the flooded forests of Cambodia, and from the frozen tundra of Siberia to the salt lakes of the Australian outback, wetlands are an in-between world. Sometimes wet and sometimes dry, sometimes land and sometimes water, sometimes saline and sometimes fresh, they change their character with the seasons. Their wealth is tied up in hard-to-measure intangibles such as flood pulses and the fertile silts suspended in their waters. They may stay virtually dry for several years before being replenished by violent floods. Their wildlife is similarly transitory, with migrating birds, fish, and even mammals coming and going. And like rainforests, they are generally not owned by anyone. Their rippling waters, shifting land, and migrating wildlife are a common resource available to anyone willing to brave the elements to get them.

These are all unhelpful attributes in the modern world, where certainty in nature is valued more than fluidity, where floods are a "bad thing," where private property is the universal currency, and where resources constantly on the move are hard to own. In every sense, the wealth of wetlands can slip through your fingers. The modern world wants to enclose and privatize and tame them, much as the water in flowing rivers is captured by damming them. The owners of wetlands, despairing of exploiting the moving feast, often prefer to drain it all, fence in the land, and start again. And for generations engineers have regarded any water draining into wetlands as somehow wasted and ripe for diversion to other human uses.

So the world's wet places are being emptied. By some estimates, half of them have gone already. Next to rainforests, they are the most productive ecosystems—for humans as well as for nature. Yet they are disappearing with far less fuss: drained, dyked, canaled, concreted over, turned into shrimp farms and rice paddies, dammed, dredged, and filled with solid waste.

As one recent international research project on wetlands economics observed in the journal *Nature,* "The social benefits of retaining wetlands, arising from sustainable hunting, angling, and trapping, greatly exceeded the agricultural gains" if the wetlands were drained. And yet they are still drained, because the social gains often accrue to people beyond the wetland, while the agricultural gains can be captured entirely by the owner of the land. The study concluded that "our relentless conversion and degradation of remaining natural habitats—including wetlands—is eroding overall human welfare for short-term private gain." As scarce natural resources become rarer, the social benefit from protecting them grows rather than diminishes. Preserving what remains "makes overwhelming economic as well as moral sense."

⁓

This is a truth that the inhabitants of wetlands—the people willing to get their feet wet to harvest the waters—have known for a long time. Take the case of the English fens, a wide expanse of low farmland stretching from the great cathedral city of Ely to the North Sea. Today it is an entirely manmade landscape. Well, almost. Fly over it and sometimes, in among the straight lines of drains and roads and field margins, you can see meandering across the black peat the ghostlike, silty remains of long-vanished riverbeds.

Until the late Middle Ages, the fens were a world of bog and marsh, wandering streams and great natural wealth. Locals set out from patches of higher land, such as the Isle of Ely, to harvest that wealth from flat-bottomed boats, casting nets in the muddy waters. Chroniclers in the twelfth century wrote of salmon and sturgeon in the waters in such quantities "as to cause astonishment in strangers, while the natives laugh at their surprise." Fast coaches took live fish and fowl and eels to London markets. Eels, after which Ely was named, were the most abundant commodity and acted as a local currency. Fenlanders paid their rent in eels, and the Domesday Book records that in 1086 the town of Wisbech paid tax amounting to 33,260 eels.

Wealthy landowners such as the Duke of Bedford eyed this common wealth with envy. But it was too mobile: hard to catch and difficult to fence in. So they determined that they could make more money more easily by draining the fens and replacing the eels and fish and wild ducks with cattle and sheep. Dutch engineers were the drainage masters of the age. They had built an entire country with drains. In the late sixteenth century, Dutch engineers began to arrive in the drawing rooms of the English gentry and the corridors of the treasury, making the pitch to be allowed to recast the finds. Soon, lavishly remunerated, they were setting up their windmills across the marshes.

The most celebrated and industrious of the Dutchmen was Cornelius Vermyuden, who dug six great artificial rivers across the marshes for the Duke of Bedford. But the fenlanders saw no advantage for themselves in this draining and fencing in of their marshes. They protested with stones, shovels, axes, and a chant: "For they do mean all fens to drain / And water overmaster, / All will be dry and we must die, / 'Cause Essex calves want pasture."

One of the champions of the fenlanders in the 1620s was Oliver Cromwell, who had inherited a fenland estate from his radical uncle. He first tasted armed rebellion here. The fenlanders helped to build his reputation. But the shifting waters of wetlands seem to breed traitors. Two decades later, having led the English revolution and beheaded King Charles, he joined forces with the Duke of Bedford in a project to drain the fens and sent his army onto the marshes to suppress the protesting commoners.

For a long time the drainage engineers' reach exceeded their grasp. In their efforts to drain the marshes, they caused drought and flood. Bankruptcies were frequent. Their great public works were constantly bailed out with public funds. And all the time the winners in the enterprise seemed less numerous than the losers. Even today, when the drains have finally succeeded in their task and fenland farmers make good profits, the public expenditure to keep the pumps running far outweighs the private gain.

The story of the fens is worth telling because it is so similar in so many ways to modern efforts around the world to tame the wetlands. Most such efforts result in huge public expenditure that wrecks natural ecosystems of staggering fecundity in order to generate private profits that are puny by comparison with the former common wealth.

Let us settle for three examples here, from three of the world's largest wetlands—the Sudd swamp on the Nile River in Sudan, the Pantanal in the heart of South America, and the Okavango delta in southern Africa. Two of these involve plans for canals that would rush waters past the wetland, in one case to reduce evaporation, in the other to aid navigation. The third is an apparently modest attempt to tinker with a virgin ecosystem for laudable development goals, but it could have huge unintended consequences. In all three cases, the potential beneficiaries are remote from the wetland and have little to lose by its demise. In each case, too, a common resource of global importance is at risk of being destroyed for narrow self-interest.

The Sudd is the world's second largest swamp. It occupies a swath of southern Sudan on the White Nile, one of two great tributaries of the Nile River. Places do not come much more remote than this, and the Sudd is a formidable natural obstacle. It prevented navigation upstream by generations of European explorers seeking one of the Holy Grails of geography, the source of the Nile.

The Sudd is a natural wonder of the world. Its numberless channels contain an ever-shifting maze of papyrus islands, the *sudds* (Arabic for blockages) that give it its name. *Sudds* can be up to half a mile long and many yards thick —thick enough to carry large herds of hippos and elephants. Sometimes they block the whole channel until, like a dam, they burst. This must be an extraordinary sight. One British colonial officer on a rare nineteenth-century foray into the swamp described how, after such a burst, "crocodiles were whirled round and round, and the river was covered with dead and dying hippos." Others were not so enthused by the Sudd. The Victorian water engineer Sir William Garstin said of it, "No one who has not seen it can have any real idea of its supreme dreariness and its utter desolation. To my mind the most barren desert is a bright and cheerful locality compared with the White Nile marshland."

The Sudd is a 300-mile turnout on the long journey of the White Nile's waters from Central Africa to the Mediterranean. The river's water takes more than a year to pass through the swamps, during which time half its volume, some 4 million acre-feet, is lost to evaporation. Those are acre-feet that people downstream have long wanted to get their hands on. It was the unenthusiastic Garstin who first proposed either dredging the Sudd or diverting the White Nile away from its suffocating embrace in order to preserve its water from the sun. He never got his way, but modern engineers have not lost the urge. Egypt, which is already using every drop of the Nile that flows through its land, wants to turn dreams into action. And that is why for the past thirty years a vast Rube Goldberg machine called a "bucketwheel" has been sitting in the middle of the bush of southern Sudan.

The bucketwheel is the canal diggers' answer to the combine harvester. It is a giant laser-guided digging wheel weighing 2500 tons and as tall as a five-story building, a fearsome parody of an ancient Persian waterwheel with twelve buckets attached to the wheel to grab earth rather than water. It was in fact constructed in the 1960s to dig a canal across the Punjab in Pakistan. After that job was done, it was dismantled and carried by truck, train, ship, and finally camel to the edge of the Sudd. Its new task was to dig a channel 160 miles long and 160 feet wide to allow the Nile to skirt the eastern side of the marshes.

In June 1978, the bucketwheel began excavating what was called the Jonglei Canal. The giant wheel rotated once every minute, night and day. At each revolution its buckets ate up and threw aside enough Nile sand to fill an Olympic swimming pool. The machine consumed 10,500 gallons of fuel a day—more than all the buildings in Juba, the regional capital.

This, however, was not a good moment to dig a canal through the backwoods of southern Sudan. The route passed through the land of Dinka cattle herders. They were seriously inconvenienced by the canal, which cut off their lands from dry-season pastures on the fringes of the Sudd. Moreover, the Dinka and other groups in the non-Muslim south of the country were about to embark on a war of liberation from their Muslim masters far to the north in Khartoum.

By February 1984 the bucketwheel had carved out a third of the route, at a cost of $100 million. Then the Sudanese People's Liberation Army struck the contractors' camp. They destroyed everything except the great machine itself and ran off with foreign hostages. This was no ignorant piece of terrorism. Its perpetrator, the leader of the SPLA, was John Garang, a young Dinka tribesman who had received an American university education in the 1970s. While at Iowa State he had written a doctoral dissertation on the iniquities of the Jonglei plan. In it, he argued that the canal would suck southern Sudan dry of its greatest resource—the waters of the Sudd.

Though the hostages were released unharmed a year after their capture, the contractors never returned to their project. Civil war raged around the Sudd for twenty years, and the bucketwheel has sat, more or less intact, beside the wetland ever since. At the end of 2004, Garang and the authorities in Khartoum concluded peace talks that provided a fair degree of autonomy for southern Sudan. Insiders said that twenty years on, Garang was better disposed toward the Jonglei project and might finally consent to the construction of the canal, with proper safeguards for his people. After his death in an air accident in 2005, it was far from clear what the consequences might be for the Jonglei. But Egypt was certainly eager to put up the cash. The new millennium saw it engaged in a giant new Saudi-funded irrigation project in the western desert, known as the Toshka Lakes scheme, which will eventually require 4 million acre-feet of water—coincidentally, precisely the anticipated yield of the Jonglei Canal.

The canal may now be dug. However, in their concern to end evaporation of Nile water from the Sudd, the Egyptians seem to have lost sight of an even bigger evaporation loss far closer to home. Though it is rarely mentioned, roughly three times more water evaporates every year from the surface of Lake Nasser, the vast reservoir behind Egypt's Aswan High Dam, than from the Sudd. This Russian-built symbol of Egyptian independence, the crowning glory of the rule of Colonel Abdel Nasser in the 1960s, loses roughly 12 million acre-feet from its huge surface every year. In a dry year, that is more than a third of the river's entire flow. But nobody is suggesting that the reservoir should be emptied.

———

The engineers' second great target is the Pantanal. The largest freshwater ecosystem in the world, it is even bigger than the Sudd. Sprawling across the huge floodplain of the Paraguay River, it spreads from Brazil into Bolivia and Paraguay in the heart of South America. Its swamps cover an area five times the size of the Netherlands and are home to huge populations of jaguars and armadillos, caimans and anteaters, giant otters and anacondas, the world's largest snake. Above them all stands the giant "too-yoo-yoo" bird, the Jabiru stork, one of more than six hundred bird species that live here.

A fifth of the wetland—the biologically richest part—has been made a World Heritage site. A recent economic assessment of its "ecological services" put its value at $16 billion a year, primarily for its regulation of water. It is like a vast natural sponge, maintaining a regular water flow downstream through Paraguay's two major cities, Asunción and Concepción, all the way to the Argentine capital, Buenos Aires.

The Pantanal is, like all such areas, being encroached on by farmers and hunters. But their effect is small in comparison with the likely impact of a plan to turn 2000 miles of the meandering and often impassable Paraguay River into a giant shipping canal to the sea. The $3 billion Hidrovia Project would cross right through the Pantanal, allowing oceangoing ships to jostle with the wildlife as they head upstream to the Brazilian inland port of Cáceres at the northern end of the wetland. There they would load up with gas brought by pipeline from Bolivia and timber and soybeans harvested in the remote Brazilian bush. The waterway would, its proposers hope, act like navigation on the Rhine in the nineteenth century, as a conduit and magnet for industrialization in all three countries. And its influence could be greater still. It might ultimately link up with other waterways proposed for the Amazon and Orinoco basins, ultimately creating a water highway across South America, from Argentina to Venezuela and Bolivia to the Atlantic.

The Hidrovia Project, like the Jonglei Canal, has been in the planning for a while. The region may be peaceful, but opposition to the scheme is still intense. Local people call it "hell's highway." The proposal was last abandoned by Brazil in 2001, but it keeps coming back. Government scientists insist that "the work in the Pantanal will be like a needle to an elephant—it will not feel a thing." But independent hydrologists say that the waterway would drain 14 million acre-feet of water out of the Pantanal and dry up half the wet areas in the dry season, when the few areas that remain flooded serve as important refuges for aquatic animals.

———

In truth, the Pantanal is not an unchanging wilderness. More than 90 percent of the land is privately owned for cattle ranching. The protected natural wetland is just a 400-square-mile core. But it is remote. The Chamacoco Indians in the southern tip of the wetland can be reached only by boat or plane. In recent years, river flows into the wetland have been heavy. Many of the ranches have flooded. And as wildlife flees from economic development in the adjoining forests and grasslands, the Pantanal has become an increasingly important refuge for South America's remaining wild creatures. Its loss would be an ecological tragedy to set beside the razing of the Amazon jungle. And its hydrological impact, in increased droughts and floods, could be of huge importance to millions of people downstream all the way to Buenos Aires.

———

Finally, there is the Okavango delta in southern Africa. This is an inland basin—the swampy endpoint for the Okavango River in the Kalahari Desert in Botswana. Its waters come from the forests of Angola, pass briefly through Namibia in an area known as the Caprivi Strip, and end up in a wide, flat, fan-shaped delta in the desert. The delta is a land of papyrus and hippo grass, lily-covered lagoons and peat islands that occasionally dry out and burn. It is home to tens of thousands of elephants, hippos, zebras, buffalos, and antelopes and has for centuries been an oasis for the tribes of northern Botswana.

Unlike many desert wetlands, where water evaporates in the blinding sun, the Okavango delta's water stays fresh. This is because the half-million tons of salt that the river brings to it each year becomes concentrated in underground water that flows into small islands, where the salt is bound to sand that is also brought down in the river. It is a remarkable process, and ecologists are afraid that any meddling with the delta's hydrology could destroy it.

Now Namibia has plans to dam the Okavango River for hydroelectricity on the Caprivi Strip, just 30 miles upstream from the delta. Angola, now freed from a generation of civil war, also wants to build dams in the headwaters of the river. Fears for the future of the Okavango are growing. Engineers say the ecologists are panicking unnecessarily. The Namibian dam would be only about 20 feet high and would hold back only a small amount of the river's flow, they say. But Terence McCarthy, of Witwatersrand University in South Africa, warns that even this dam, which is really little more than a weir, would block the river in the dry season, preventing around 80 acre-feet of sand from being washed into the delta annually. That sand, he says, is critical to the process of quarantining the salt. Without the sand, the salt will spread through the delta, destroying its complex ecosystems one by one. Not only would that be a wild-life tragedy, it would also destroy one of the linchpins of the Botswana economy—wildlife tourism, which is concentrated in the delta.

Will the bucketwheel finally carve its way across the savanna of southern Sudan, draining one of the continent's most fecund environments to fill the irrigation channels of Egypt? Will the hidden wealth of the Pantanal slip through South American fingers like so much silt? Could dams doom the Okavango? We can make the moral and the environmental cases for saving these oases. But much may depend in the end on the abstractions of economists and whether they can come to recognize that commonly owned wealth has as much value and importance to society as privately owned wealth.

10

Lake Chad:
Tragedy of the Floodplains

The death of a wetland is a terrible thing, particularly a wetland in a desert. When it happens, lakes shrivel, crops go brown in the baking sun, fishing nets empty, trees die, and herders slaughter their animals for whatever pitiful amount of cash they can raise. The land curls up and dies. The people depart. Usually this happens when the rains fail. How much more terrible when the drought is manmade—when the wetland dies because humans have decided to divert the rivers that should replenish it; when the water is taken for little purpose, as a statement of the power of one community over another; and when even hard-nosed economists say that what is going on is madness.

Welcome, then, to the death of the Hadejia-Nguru wetland in northern Nigeria. This dramatic, kite-shaped expanse of green and blue stretching for more than 60 miles along the edge of the Sahara Desert was once a bulwark against the advancing sands. Its lakes were full of fish, and their fertile waters washed annually across the land, creating lush pastures that sustained tens of thousands of cattle and watering more than half a million acres of fields that farmers planted with crops as the floodwaters receded. The wetland supported a million or more people and provided exports of fish and vegetables to cities across the largest country in Africa.

Some of this remains. But much has been destroyed as dams divert water upstream for irrigation projects, the lakes shrink, and the summer flood dries up. In a region of Africa known for drought and famine, the government promised that the dams and irrigation works would turn the landscape green

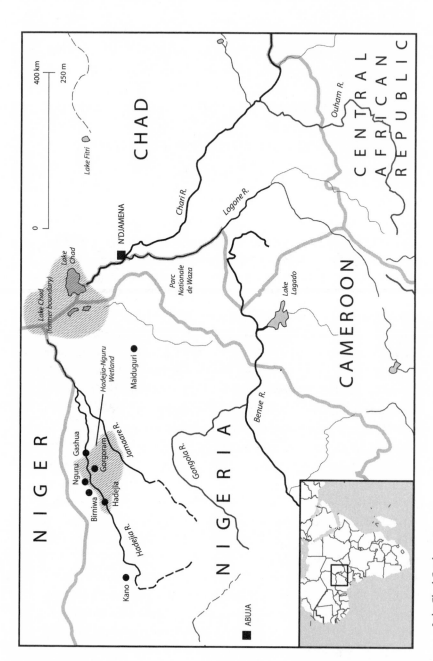

Lake Chad Catchment

and create bounty. But they have been doing the opposite. They have been destroying a natural bounty of far greater value.

Westerners are used to keeping rivers within narrow channels. For us, floods are generally a bad thing. When African engineers train in Western colleges, they learn this viewpoint and go home to banish floods. But in Africa and much of the developing world, poor farmers and fishermen and cattle herders need floods. They bring free irrigation water and silt to fertilize fields and sustain pastures and fisheries. When engineers capture the flood flows, the millions who depend on the floods suffer.

The Hadejia-Nguru wetland is a complex network of natural channels and lakes that mingle the waters of the Hadejia River, coming in from the west, and the Jama'are, coming from the south. But two dams have been built on tributaries of the Hadejia. Between them, they capture 80 percent of that river's flow. The area of floodplain regularly inundated has shrunk by two thirds. Lakes like the Punjama and Nguru have also shrunk, and the water table has declined by as much as 80 feet, unleashing a further round of desiccation. Meanwhile, on the Jama'are River sits the half-completed Kafin Zaki Dam. The money for this ran out long ago, but the government still wants to finish it. If that happens, Kafin Zaki will be the largest dam on the system and will dry out the remainder of the wetland for good.

A journey across the wetland shows the hydrological devastation and the tensions the dams are creating. On the east side, Gorgoram was once a sizable town in the heart of the wetland. It had a famous fishing festival, to which teams of youths came from hundreds of miles around to net the biggest catches in the last pools as the flood retreated. Dignitaries from across northern Nigeria came too, awarding prizes and lobbying for votes. But now Gorgoram is an emptying village. The festival is failing, because the floods are poor and there are no fish. And all around the landscape is littered with fallen trees, victims of the falling water table.

Many villages have lost their wells and lakes and woodlands. At Adiani, on the southern shore of Lake Nguru, villagers told me how they once lived by keeping bees in the village's tiny forest reserve. But the forest is dying, and with it the bees. On the road to Nguru town, I drove through an empty wasteland that went on for miles. It seemed deserted until suddenly a man appeared from behind one of the last trees left standing. He was like an apparition, and probably sensed it. He advanced on our car with rage in his eyes. "This is how men are suffering in the bush," he roared, and would say no more.

On Lake Punjama the waters were still high, and hundreds of birds were swooping low. Locals took me fishing in a flat-bottomed boat. We landed some catfish. In the soft early-evening light, we rowed past boys floating on large pumpkin shells. They had knives to cut the tall grasses to use as floats for their fishing nets. And nearer the shore, two brothers were wading waist-deep in the water, driving their cattle to the same grasses to feed. The first group shouted to the second to keep the animals clear of their fishing lines, which we saw paid out across the water.

It appeared an idyllic scene. But an hour later I met a family of Fulani cattle herders setting up camp for the night on an island in the lake, and they had a different story. Like other Fulani, they came to the wetland during the dry season to find water and graze their animals on the lake edges and on stubble in fields after the harvest. The woman rounded up the cattle, shooed them into a stockade, and lit a fire of dung and straw to keep away the mosquitoes. But her husband was morose. "Once many Fulani were coming here," he said. "There was more water then, and more grazing land. Now there is little grazing. We still come, because there is nowhere else to go, but there are always disputes with the farmers." The previous year there had been pitched battles between the Fulani and gangs of settled farmers, he said. By the time the fighting died down, bodies were floating in the water and ten people were dead. "Every year, people get killed."

The first dam was built on the Hadejia in 1974. The Tiga Dam stored water to supply a state irrigation project south of Kano, the biggest city in northern Nigeria. Each wet season, the rain must replenish the reservoir and supply the irrigation project before any surplus can flow down into the wetland. In the early 1990s, the government completed the Challawa Gorge Dam on another Hadejia tributary. It was built to supply water to the 30,000-acre Hadejia Valley irrigation project on the edge of the wetland.

This scheme has turned into farce. For more than a decade French engineers formed a garrison in the wetland, surrounded by earth-moving equipment and waiting to build the $100 million project. But the money kept running out. When I visited, huge areas of forest had been cleared to make way for the irrigation channels. I spoke to the emir of a nearby village, who hoped his people could move onto the land within a year or so. But after years of promises, the only crops growing here were in the French project director's vegetable garden (I had some of them for lunch). Today less than half the project has been completed. Soils that had nourished a forest before the trees were chopped down have baked hard. Some believe that the whole project is doomed. Some just hope so.

———

Since the completion of the dams, the water that has come down the rivers and onto the wetland has often caused mayhem. It has come at the wrong time, fed the wrong channels, and caused damaging floods rather than nurturing the ecosystem. Take the 40-mile-long Burum Gana Channel. Once it was one among many that distributed the Hadejia's waters. It was wet for just a few months a year. But it is now the only channel that gets any water, and it is wet all year. The perennial flow has attracted bulrushes, which now choke the channel, often blocking it entirely so the water overflows and destroys crops in surrounding fields. Some communities have been forced by the floods to abandon their villages. Birniwa, which used to be among the more parched parts of the wetland, is now flooded almost every year. Meanwhile the bulrushes are invading fields where rice once grew. And the fields have attracted a huge invasion of quelea birds, which roost among the bulrushes and forage for crops all across the wetland. According to local press reports, in some villages children are kept home from school to frighten the birds away.

Occasionally there are bigger floods, when, after very heavy rains, dam managers hurriedly release water downstream to protect their structures. In August 2001, 140 people died and thousands lost their homes as a wall of water rushed across the parched wetland.

Is this hydrological havoc the price that has to be paid for economic progress? Well, no. Economists have concluded that the big dams and irrigation canals here have been a huge waste of money. The British government's aid agency calculates that the income from each acre of land has fallen since the dams were built from $68 to $8. The value of the crops grown on the irrigation projects, it says, is tiny compared with the lost benefits on the wetland from farming, fishing, harvesting timber, raising livestock, gathering honey, cutting straw for construction, and making bricks.

Ed Barbier, an environmental economist from the University of Wyoming, agrees: "Gains in irrigation values account for at most around 17 percent of the resulting losses on the floodplain." There is, Barbier says, "a clear choice between using the water for irrigation upstream and allowing it to enter the wetland to support agriculture, fishing, and other economic activities there." And in economic terms, he says, there is no contest. The wetland wins every time. The half-completed Hadejia Valley Project should be abandoned, along with the mothballed Kafin Zaki Dam on the Jama'are. The only sensible use for the existing dams, he says, is to recreate the natural flood. But the bigwigs in Nigeria have yet to take such advice.

I moved on to Gashua, the easternmost town on the Hadejia-Nguru wetland. It is where the various channels coalesce again to form a new river, the Yobe, which heads for Lake Chad, an inland sea in the heart of the Sahel. These days only a quarter as much water emerges from the wetland as before the dams were built. In Gashua, fish were in short supply in the town market, owing to the declining productivity of the wetland and the river. What the town's traders did have, I discovered, was bags of potash—the salty crust of desert soils, scraped by villagers from shriveling oases to the north. It was the one local commodity in plentiful supply.

I took a detour to one of these oases, Garin Momadu. Even here there was desolation. The wells at the oasis were running dry, dead palm trees lay across the ground, and sand dunes were taking over the houses on the edge of the village. Heading back to Gashua, we passed a camel train—a dozen animals all bearing yet more potash to the market. The animals had just crossed one of the beds of the Yobe, which hydrologists say feeds the oases. But it had been dry, locals said, ever since the Tiga Dam was finished. That probably explained the crisis at Garin Momadu. The bed was just a sand-filled depression, and the air was full of dust whipped up by the desert winds. Camels walked where fishermen once threw their nets. Ahead of the camel train a man strode alone in white robes, carrying a sword. The desert was on the march.

On again, to Maiduguri, a town with another ominous warning of what awaits the wetland. The largest town in the area, it had already exhausted two aquifers and was pumping from a third. Officials from a European Union anti-desertification project in the town told me that without river flooding, this final aquifer would never recharge. The town had built a reservoir to take water directly from the river. But the river was nearly dry, and the reservoir lost more than 6 feet of water a year from its surface to evaporation. The town would die unless another source of water could be found.

———

Does it have to be like this? Witness events on another river that drains into Lake Chad, the Logone, in the neighboring state of Cameroon. The Logone is bigger than the Hadejia ever was, because it flows out of the rain-soaked fringes of the great Congo rainforest of Central Africa. Its waters have for centuries been the lifeblood of a thriving floodplain rich in wildlife and with ample pickings for fishermen and hunters. Farmers planted traditional varieties of rice on the wet soils left by the retreating floods each year. The Logone also attracted Fulani nomads and their cattle from neighboring countries. It is said to have been one of the great centers of the Fulani, the largest nomadic pastoral group in the world.

But in 1979 a state-owned rice company built the Maga Dam to capture water entering the Logone from the Mandara Mountains, to the west. The idea was to divert that water to irrigate new fields of modern rice varieties on the floodplain. Modern green-revolution rice is more susceptible to excess water than traditional varieties, so to prevent possible flooding from the river, the company raised 60 miles of embankments to separate the river from the paddy on its old floodplain. The result was to deprive the river of water and wreck the floodplain ecosystem. A hundred thousand people who had depended on the floods for their livelihoods were left destitute.

Paul Loth, of the World Conservation Union, conducted a comprehensive assessment of this project and its aftermath. He concluded that it was spectacularly inefficient even in its own terms. Little rice was grown most years, and by some measures only around a tenth of the water collected by the dam has ever been productively used. Like the irrigation projects in northern Nigeria, it "diminished rather than improved the living standards and economy of the region as a whole. A once fertile floodplain was turned into a dust bowl."

The litany of disasters assembled by Loth is terrifying. Rich pastures of perennial grasses died, so that some 20,000 head of cattle had to move away. Fish yields fell by 90 percent. Water tables fell for more than 600 square miles, emptying wells and water holes. Yields of the local varieties of sorghum and rice fell by 75 percent. Meanwhile, elephants and lions in the Waza National Park, one of their last refuges in Central and West Africa, fled as their water holes dried up. Those that stayed were mostly killed by desperate farmers, who moved in to poach them. With the entire floodplain in crisis, the human inhabitants fought over water and pastures. Many left for distant cities.

But, unlike in Nigeria, that was not the end of the matter. Local leaders joined foreign scientists to implore the rice company to allow water back onto the floodplain. They argued that it could be done without destroying the irrigation project. The dam managers briefly relented. In 1994 and again in 1997, they let water downstream from the dam, flooding about a third of the former floodplain. Zanankay Yap, a traditional leader in the town of Tekeleon the floodplain, remembered when the embankment was broken. "We all waited to see the water arrive with our own eyes," he said. "And then when it came, there was instant joy. We farmed well and we fished well that year."

The pilot releases had a lasting effect. The perennial grasses returned. Cattle numbers increased threefold. Fish bred again in the pools. Wildlife began to return to the park. Wells filled, and the greater availability of clean drinking water improved local health. Economists from the World Conservation Union measured the short-term economic benefit and used the results to calculate what had been lost by having the dam dry up the floodplain. The dam and embankments, they said, had cost eight thousand of the poorest and most vulnerable households in this part of Africa a total of $2 million every year for over twenty years. That is almost a dollar a head per day and far greater than the gains from the expensively watered rice paddies. "Often nature is a triple-A company, and its dividends go to its shareholders—the rural poor," said the economists.

The lesson is obvious. The water should be given back to those who need it most, and who evidently use it most effectively—the residents of the floodplain. And yet, extraordinarily, those pilot releases—at the time of writing, almost a decade old—have never been repeated. The embankments that cut the river off from its floodplain have been put back. In March 2004, after several years of pushing for a permanent resumption of the flooding, the scientists from the World Conservation Union packed their bags and left.

It is too easy to see communities that depend on natural wild resources and the vagaries of untamed rivers as somehow left behind by progress. The truth, quite often, is the opposite. It is they who have unlocked the truth about how to make the maximum use of natural resources. It is the urban sophisticates with their engineering degrees who haven't got a clue. The pilot studies on the Logone and similar research done on the Hadejia-Nguru wetland cement the case for a completely different way of managing floodplains and the rivers that sustain them, not just on the fringes of the Sahara but right across Africa, where millions still depend on their free water services.

———

What does all this mean for Lake Chad, the great inland sea on the edge of the Sahara into which the Yobe and the Logone, along with other rivers, like the Chari, drain? The lake drains an area as big as France, Spain, Germany, and the United Kingdom put together, and atlases show it as having a shoreline in four countries: Nigeria, Niger, Chad, and Cameroon. But it is naturally temperamental, given to huge changes in size. This is partly because the rivers themselves are temperamental, even without human influence. And it is partly because the lake is so shallow, currently averaging only 5 feet in depth. It is thus very prone to water loss from irrigation, and very dependent on the rivers for renewal. Ten thousand years ago, the lake filled its basin and overflowed through Nigeria into the Atlantic Ocean. And yet it entirely dried up four times between 1400 and 1910, when it divided into two small pools.

After 1910 it grew again, reaching a peak in 1962, when it covered almost 10,000 square miles. Since then it has been in retreat once more. By the late 1990s it had lost 95 percent of its surface area. The open water had again been divided in two, with a large swamp in the middle. By 2004 the lake had shrunk again and was reportedly down to 200 square miles. All this happened because in forty years, the amount of water coming down the rivers has halved. Hydrologists figure that between a half and two thirds of this decline is due to diminishing rainfall and the rest to diversions for the irrigation projects in Nigeria and Cameroon.

The retreat has caused enormous disruption. Niger and Nigeria no longer have a lake shoreline. Nigeria and Cameroon have long disputed where exactly in the lake the border between them runs, but it only mattered after the lake retreated and farmers moved onto the disputed bed. The International Court of Justice finally ruled that Darak and thirty-two other villages were part of Cameroon.

Worse, government infrastructure and Western aid projects have everywhere proved quite unable to cope with the lake's decline. Three decades ago, British engineers began to build what should have been one of Africa's greatest irrigation schemes, the South Chad Irrigation Project. The project was intended to irrigate rice and wheat on a vast 165,000-acre stretch of the lake's southwestern shore. There was a 19-mile intake canal from the lake. But even that wasn't enough to keep up with the retreating waterline. Most of the project area has spent most of the time since then high and dry. Its grain mills are rapidly becoming of interest only to industrial archaeologists.

But the local farming and fishing communities, who generally have little to do with central government and rely on their traditional systems of rule, have taken hydrological events much more in their stride. The British social economist Terri Sarch has sent several years visiting the lake. She says that while the South Chad Irrigation Project sat abandoned, the farmers who were supposed to benefit from it simply moved with the lakeshore to cultivate the lakebed. There were no roads, no government wells, and no police—but there was wet, fertile land. Roughly a million acres of it. And that was far more important.

Interestingly, the constant movement and establishment of new farms on new land has been a remarkably peaceful process. Sarch says the farmers fell back on their traditional systems of authority for allocating land. A typical community, from the Nigerian village of Tumbun Naira, moved east in 1985 to farm on the lake floor close to an abandoned fishing camp. They stayed a few years before moving again, in 1994, after rising lake levels flooded their settlement. But they returned and rebuilt the following year, as the lake retreated once more. It sounds a like a recipe for conflict, failed crops, and famine. Yet "they did better than merely cope," Sarch says. "Many are producing a food surplus." The evidence included an influx of pickup trucks driving across the lakebed.

The lesson of all this? When the rivers run dry, it does not need to be a disaster, provided societies can adapt to cope with it. And the traditional attributes of flexibility associated with communities living on wetlands serve remarkably well. One of the ironies is that we have grown disturbingly good at disrupting river flows while losing our capacity for coping with, let alone prospering from, the consequences.

11

Seas of Death

For thousands of years, under a dozen different empires, the Hamoun wetland has been a refuge and a source of food and water in the fierce deserts of Central Asia on the Silk Road to China. Overlooked by the mysterious Parthian citadel of Kuh-e Khwaja, the Sistani people of the wetland, on the remote border between Afghanistan and Iran were always left to their own devices. Or so it was until a lethal combination of drought, U.S. engineering, and Taliban fanaticism turned their oasis to dust at the start of the new millennium.

The Hamoun wetland, as conventionally mapped, covers 1500 square miles. Even in bad years, half that area would be wet. Leopards lurked in the marshes, and carp and otters swam in the three lakes at their heart. It was a mecca for flamingos, ducks, and pelicans migrating from Siberia to the Indian Ocean—and, of course, for the Sistani, who lived apart, fishing, hunting, farming, and punting across the lakes in traditional flat-bottomed reed boats called *tutans*. They herded their animals across the rich pastures around the lakes and tapped the waters to irrigate wheat and barley, grapes and melons, even a little cotton and sugarcane. As recently as the mid-1990s, they pulled thousands of tons of fish from the lakes each year.

The water came from the snowfields of the Hindu Kush in the east of Afghanistan. The Helmand River flowed for 800 miles across Afghanistan, through the Margo Desert, and into the Sistan depression, which straddles the border with Iran. There its water slowly evaporated in three lakes, each usually not more than a couple of feet deep. The upper lake, which got water first, was in Afghanistan; the two lower lakes were in Iran. Unusually in such an environment, they each contained fresh water, because the occasional years of floods overflowed the lakes and flushed the salt into salt flats to the south.

But all that disappeared in 1998, when all three lakes dried out. The reed beds vanished, the wet marshes were replaced with salt flats, and the wildlife largely disappeared. It was a major environmental catastrophe, more complete and devastating even than the destruction of the Aral Sea 600 miles to the north. And for the wetland's quarter-million inhabitants, it was a human catastrophe. But you might not have read about it, for the wetland was on a remote border between a pariah state then run by warlords and another run by a member of George Bush's "axis of evil." Only satellite images of a new desert and a few voices from refugee camps have revealed what has happened.

The origins of the tragedy go back to the repeated efforts by Afghans and their Western friends over more than a century to modernize agriculture by irrigating fields along the Helmand. From the start, the risks to the Hamoun wetland were understood. As early as 1926, under a British-brokered deal, Afghanistan promised to leave half the river's water for the Iranians and the lakes. But the deal has rarely been honored. Upstream Afghanistan has regularly taken more—especially in dry years, when the downstream need is greatest.

After the Second World War, U.S. engineers arrived in Kabul intent on transforming the Helmand Valley. They began digging dams and irrigation channels and installing hydroelectric turbines. The centerpiece was the 300-foot-high Kajaki Dam, which captured most of the river's flow in a 30-mile reservoir. Half a century on, it remains the largest and most ambitious civil engineering project ever undertaken in modern Afghanistan. It is also a model of how things should not be done.

While the hydroelectric dam at Kajaki lit the lights of Kandahar, the irrigation system proved a disaster. Less than a third of the canals intended to irrigate half a million acres of the Helmand Valley were ever built. Many of the nomads brought down from the hills to farm the desert soon returned home. But the canals that were dug consumed huge amounts of water, most of which eventually seeped into the ground or evaporated in the sun. The new canals took six times more water to irrigate an acre of land than had traditional systems, which the new canals often replaced.

Under a deal struck in 1973, the Afghans agreed to leave at least 400 billion acre-feet of water in the river each year. Various estimates put this at between a fifth and a third of the long-term average flow. The Hamoun lakes diminished but survived. But when the Taliban seized control of Afghanistan in the early 1990s, all the treaties were off. In 1998 the Taliban closed the sluice gates on the Kajaki Dam. Thereafter, no water was allowed to flow down the river into Iran until the reservoir was full. It was an act of extreme hydrological provocation. Iran protested to the UN, but nobody was listening.

Matters were made worse when 1998 proved to be the start of a drought that continued into early 2005. The combination of drought and the Taliban proved terminal for the lakes and the wetland. With rainfall at only a quarter of the norm and no flow in the Helmand, all three lakes dried out for the first time in recorded history. The marshes turned to dust bowls. Hundreds of villages on either side of the border were overwhelmed by shifting sand dunes and ravaged by summer dust storms known locally as the "wind of 120 days." The old irrigation channels around the lakes disappeared beneath the sand. "Many who had lived around the Hamoun for generations moved away or lost everything," said Hassan Partow, a researcher for the UN Environment Program, who made a rare visit.

The Sistan depression became a humanitarian disaster zone. An estimated 300,000 refugees crowded into camps there in 2002. Some had been made destitute by the environmental catastrophe; others were fleeing from the war in Afghanistan to an area where, in past crises, their ancestors had always found refuge. With unpleasant irony, some of the trucks ferrying water to the camps filled up at the same distant reservoirs that emptied the rivers in the first place.

The removal of the Taliban in Afghanistan changed little. In late 2002, the new Afghan government announced that "to show goodwill," it would open the Kajaki sluices. Some muddy water from rains in the Afghan mountains flowed across the border. Iranian peasants rushed to divert the water into their wells for storage. But after two weeks the river dried up again. "It is still dry, and there is still no agreement between Afghanistan and Iran on sharing the waters of the Helmand," Partow told me in late 2004.

Kabul claimed in its defense that it no longer had control of the river. Many Afghan farmers had built their own dams to divert any water that passed through the Kajaki Dam. Ministers added darkly that Iranian farmers were using any water they got to plant opium poppies. If so, they were not alone. In

Afghanistan, the Helmand Valley, still an outpost of Taliban insurgency, had become a major center of poppy cultivation. In retaliation, the Iranians took to blaming U.S. "provocation" for the failure of the river to flow. It wasn't clear what they meant. But the anger can only have been exacerbated when the United States announced plans to raise the height of the Kajaki Dam as part of its aid work to "rebuild Afghanistan," and when the World Food Program began feeding farmers in the Helmand Valley in return for their labor in cleaning out abandoned irrigation channels.

In the spring of 2005, nature intervened. Heavy rains far away in eastern Afghanistan broke dams and drowned an estimated two hundred people, but finally brought water roaring down the Helmund into the parched Hamoun. Two of the lakes reflooded. But as engineers rushed to rebuild the Afghan dams, it was not clear how long the reprieve would last. There was still no breakthrough in the UN's efforts to broker a deal over the river.

12

Mekong:
Feel the Pulse

The diminutive king and queen of Cambodia appeared on the balcony of the royal palace overlooking the Tonle Sap River. Like fairy-tale monarchs, Norodom Sihanouk and his queen, Monineath, serenaded their subjects—upwards of a million of them—gathered on the riverside promenade of the beautiful old colonial capital of Phnom Penh. It was the high spot of the Cambodian year, and the audience between monarch and subjects was part of one of the world's oldest boat-racing festivals, the Bon Om Touk.

That year, 2003, around four hundred boats, each decorated with paintings of water serpents, rushed two by two down the river to a finish line right by the royal palace, where the Tonle Sap flows into the mighty Mekong. Some boats contained seventy frenetic oarsmen, all standing in a line, pounding the water from the narrow vessel. Altogether, thousands took part in the races, while a million more Cambodians, from across one of the poorest and least urbanized nations on earth, flooded into the city for a weekend of eating and camping on the riverside.

This festival has taken place since the twelfth century, always at the full moon in late October or early November. It is a celebration of one extraordinary fact about the river on which it takes place. The Tonle Sap is one of the few rivers in the world that reverses its flow. It does it every year, right in front of the palace.

It happens because every year, at the start of the monsoon rains, the often quiet Mekong becomes a raging torrent that overwhelms its tributary, forcing

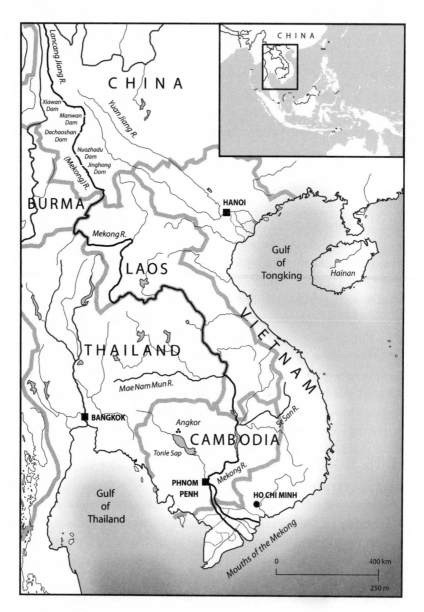

Mekong River

it to surge back upstream for some 125 miles. At the height of the monsoon, this reverse flow swallows a fifth of the Mekong's prodigious waters, and the tiny Tonle Sap is for a few months one of the world's great rivers—albeit flowing backward. The river backs up into a great lake that itself spills over its banks, flooding a wide area of rainforest, which turns into a vast fish nursery for the Mekong in the heart of Cambodia.

For five months fish grow in profusion in the silty forest waters, before being washed back into the main river as the monsoon abates and the Tonle Sap resumes its proper flow. The annual boating festival marks this unique spectacle and celebrates the great flush of fish that leaves the forest with the water. It is a flush that, in the months to come, will feed tens of millions of people up and down the Mekong River system.

Or that is how it should be. But in 2003 there was consternation on the river. The Mekong floods had been the poorest on record, the reversal of the Tonle Sap's flow had been short and tentative, and the fishing had been terrible. Some in the crowd said that mysterious changes far upstream on the Mekong had neutralized the flood. Fears grew that the reversal of the Tonle Sap might be about to come to an end. Something else was afoot too. In the royal palace. This turned out to be the eighty-year-old king's last serenade of his people. In failing health, he abdicated a few months later. Was this, his subjects asked, the end of an era? The end even of the great flood?

———

Once the world's rivers teemed with fish. Then, during the twentieth century, most of the rivers were barricaded by dams and their wild flows were tamed. Almost everywhere this has caused a drastic decline in wild fisheries. Natural fish nurseries have been wrecked and fish migrations disrupted by the unnatural flow rhythms created by the dams. But on the Mekong, the great artery of Southeast Asia, half a century of warfare has kept the dam builders away, and the natural seasons of the river have continued uninterrupted. The river has remained free to trespass over its banks and flood forests in summer, then retreat to small pools and rapids in winter. The fish have prospered. And not just the fish, for the fishing communities have enjoyed a bonanza now extremely rare in the world. In Cambodia, where the river flood is most intense, the people live off wild freshwater fish to an extent unknown anywhere else. Even the poorest can dine like kings.

The Mekong remains true to its popular image as the "sweet serpent" of Southeast Asia. It winds for 2800 miles out of the ice fields of eastern Tibet and through a long series of deep gorges in the mountains of southern China before tumbling down rapids to flood the rainforests of Laos and Cambodia and sliding into the sea through its delta in Vietnam. The Mekong is far from being the world's largest river. Its average annual discharge of 380 million acre-feet makes it fourteenth in the riverine pecking order. But its flow is today the most variable of those of the major world rivers. During the summer monsoon, it contains up to fifty times more water than in the long dry season. Then it has the third biggest flow of any river in the world, exceeded only by the Amazon and the Brahmaputra.

"The Mekong is not just another river," says Chris Barlow, a fisheries researcher at the Mekong River Commission, an intergovernmental science agency. "It is the least modified of all the major rivers in the world. Animals have evolved to exploit its flood pulse, and local societies have developed that way, too." Four fifths of the population of Cambodia is involved in fishing and processing the harvest. Some 60 million people in the Mekong's lower basin—in Laos, Cambodia, and Vietnam—draw their food and income from the river and its wetlands, catching some 2 million tons of fish a year. That is over 2 percent of the entire world catch of wild fish from both rivers and the sea. Among rivers, only the Amazon produces more.

The Mekong is a salutary lesson in what the world has lost since it began erecting concrete barriers across nature's finest. More than a thousand species of fish live in the Mekong, again more than anywhere except the Amazon. Hundreds of species are regularly caught and eaten, but the most important is the humble trey riel, a small sardinelike relative of the carp that turns up in almost every fishing net on the river. It is, says Joern Kristensen, the head of the river commission, "both meat and milk," because it provides both protein and calcium. Oxfam says that the river fisheries of Cambodia "make a bigger contribution to economic well-being and food security than in any other country." Cambodians are more dependent on wild protein than the people of almost any nation on earth. As a result, one of the world's poorest countries is one of its best fed.

The biological heart of this extraordinarily fecund river lies in the upper reaches of its reversing tributary, the Tonle Sap. In particular, it lies where the river opens out into the Great Lake, the largest body of freshwater in Southeast Asia. Every monsoon season, as the Tonle Sap backs up, the lake floods rainforest for up to 25 miles around.

After the boat festival, I took to the water: up the Tonle Sap, though the Great Lake, and into the flooded forest. It is a topsy-turvy world. Giant trees poke from the water. Tens of thousands of people live in floating villages that loosen their moorings and cruise slowly across the lake. Where there are no trees, paddy fields lie beneath the water, waiting to emerge in the dry season. One French author described Cambodia as the land where fish grow on trees, and that, more or less literally, is what most of them do—except for the striped snakehead, which lives among tree roots in swamps and can slither overland when the swamps dry up.

This flooded forest is one of the most productive natural ecosystems in the world. Each summer billions of fish fry are swept here on the monsoon surge to feed on the floating vegetation. Then, as the forest drains each autumn, the fattened fish leave. And the local fishermen are ready. Much of the flooded forest is in private hands, divided into five hundred fishing lots that are auctioned by the government. Many are managed by Vietnamese, who are regarded as the master fishers. The biggest lots can cover hundreds of square miles and make their owners $2 million a year. From October to March, virtually the entire lake is ringed by bamboo "fences" erected by the lot managers to trap fish as they swim out of the forest.

The forest is also a magnet for thousands of Cambodia's poorest people, who take their nets into the open-access areas outside the private lots. This is currently one of the fastest-growing rural areas in the world. In the mid-1990s, there were only 350,000 people around the lake; today there are a million. Many are landless refugees from the great expulsions by the Pol Pot regime in the 1970s. They depend for survival on the free resources of the lake.

Some say all this activity is destroying the fishery. I clambered aboard a floating wooden platform cum office and met Sreng Sokhak, a fishing-lot owner with twenty employees. "I've been fishing on this river for twenty years. We used to catch a hundred tons a year. We get nothing like that now," he said. He complained about his neighbors, and about unlicensed fishermen who come in with batteries and electrodes and send an electric charge through the water to electrocute the fish. But others argue that a much greater danger would be the loss of the forest and waters that make the system so productive. They say that ultimately the fishermen are the true friends of the fish, because they prevent farmers from felling the flooded forest and converting it to rice paddies.

Whatever the truth in this debate, the fecundity of the ecosystem is extraordinary. During their time in the flooded forest, the fish are the main food for a cornucopia of wildlife that in turn becomes a target for human hunters. Gangs of villagers collect the eggs and chicks of cormorants, pelicans, storks, and ibises. Others hunt squirrels and rats, pythons and cobras, turtles and lizards. Up to a million water snakes are netted each year.

Conservationists are here, too, policing the various protected areas. Sleeping in a treehouse above the water in the Prek Toal nature reserve, I met a pair of local villagers employed by the Wildlife Conservation Society in New York. Their job was to stay out in the flooded forest for a week at a time to protect migrating birds from hunters. Their charges that week included a magnificent fish eagle, hundreds of Indian cormorants, and forty pairs of greater adjutants, which are considered to be among the world's rarest storks and breed only here and in the Indian state of Assam.

Traders come to the lake from Thailand, Vietnam, and even China to buy crocodiles, fish, snakes, and monkeys. A Vietnamese monkey farm was in the market for macaques to breed for Western laboratories. Crocodiles have been hunted to extinction in the lake and forest, but the floating villages now farm them. Stepping unawares out of my boat onto a jetty in Kaoh Chiveang village, northwest of the lake, I was surprised by a rush of water beneath my feet. Looking down, I saw between the wooden slats a score or so of young crocodiles, each about 3 feet long, in a cage suspended in the water.

The village was an upmarket suburb on the lake. The floating homes were immaculate, with flower and vegetable gardens at their doors, electricity generators hooked up to long lines of car batteries, and TV aerials on every palm-frond roof. The waterways were lined with stores, karaoke bars, filling stations, and elaborate cranes for lifting boats out of the water for repairs. I soon found out the source of the wealth. The village of some three thousand people had at least as many crocodiles. Most fed on water snakes harvested from the flooded forest all around. A fully grown reptile fetched about $1000. Every household with the capital to spare had set up a crocodile cage. The open water might now be empty of the beasts, but the cages probably contain more than ever lived in the lake.

While people working the lake today live a life afloat, their ancestors had a much more sedentary existence. On the lake's northern shore sit a series of truly spectacular temples, headed by Angkor Wat. They are the remains of the capital of the great Khmer Empire, which lorded it over the flooded forest, the Great Lake, the Tonle Sap, the lower Mekong basin, and much of Southeast Asia a thousand years ago. Beside the lake, they created what historians believe was probably the largest urban area on the planet, becoming rich on the fish of the lake and the rice grown on its shores. The thousands of carvings depicting fish and fishing that festoon the walls of the temples are testimony to the huge importance of the lake to the wealth of the empire, and to the enduring productivity of the ecosystem here.

———

Just as Angkor once ruled much of Southeast Asia, so the flooded forest is an ecological dynamo of importance far beyond its immediate surroundings. It feeds the whole ecosystem of the Mekong basin. Two thirds of all the Mekong's fish are nurtured in and around the Great Lake. The waxing and waning of the lake is, say ecologists, the "beating heart" of the river and its fishery—sucking in silt, larvae, and water and expelling grown fish a few months later. "If the heart stops, the system dies," says Anders Poulsen, a fisheries biologist at the river commission.

And the artery to the heart is the Tonle Sap. During the five months when the lake drains, the river is usually full of activity. Organized bands of fishermen suspend wooden scaffolds over the Tonle Sap and suspend from them bag-shaped wicker nets, each about 80 feet wide and 400 feet long. These "bag nets" are like a fixed trawl, in which the water rather than the boat does the moving. They can fill with a quarter ton of fish in fifteen minutes. At the height of the season, more than 30 tons of fish can be caught every hour in the Tonle Sap, day and night.

Just occasionally the bag nets bulge with a giant catfish. This is the largest freshwater fish in the world. It grows up to 10 feet long and can weigh as much as a cow. It migrates more than 600 miles up and down the river, from the Great Lake into Laos, where kings once sacrificed a man and a woman before daring to hunt for catfish. These days the giant fish is a protected species. In Phnom Penh, the bag-net fishermen reach for their mobile phones when they catch one and call Zeb Hogan. He is a biologist from the University of California at Davis but spends part of each year in Phnom Penh on catfish alert—ready to measure, tag, and release them whenever they get caught in the bag nets. The day we met, he had been up at 2 A.M. recording an 8-foot monster. They are still revered, he said. Even at that hour, women came out to sprinkle perfume onto the fish in an act of homage before their husbands released it into the water.

But the end of 2003 and early 2004 was a desperate time on the Tonle Sap. The summer flood had been poor. The reversal of the river into the Great Lake started late and finished early. A five-month reversal had become a three-month reversal. Less forest was flooded, and the fish had less time to mature. The bag nets caught a mere 6600 tons—less than half the usual haul and the worst on record. Nguyn Van Xia, a riverside buyer for a wholesale market in Phnom Penh, stopped to talk on a landing point just a hundred yards from one of the main bag-net fishing spots. It ought to have been the height of the season, but he had plenty of time. There should be twenty truckloads of fish leaving this spot every day, he said. But the current average was five trucks. Several of his friends were so short of work they were playing volleyball on the beach.

Why? He shrugged. "The flood was bad this year, so we have a bad catch," he explained. Out on the river, most of the fishermen said their catches had never been so poor. Most blamed low flows. One, heading back to his floating village across the lake with empty nets, told me simply, "When the water is shallow in front of the royal palace, there are no fish in the river."

———

The Mekong is weird. For one thing, the pulse of life on the river seems to be mediated by the moon. The trey riel, the most common fish in the river, times its long migrations with phases of the moon. And then there are the hundreds of Naga fireballs, which rise from the river during the October full moon. Even sophisticated townies are spooked by these small orbs of red, pink, and orange light, each about the size of a tennis ball. They are named after a giant mythical serpent, Naga, who is said to release them from her lair in caves beneath the river.

The balls are a genuine natural phenomenon. One explanation is that pockets of methane are drawn from rotting vegetation on the riverbed by the moon's gravitational pull. But though there may be no giant serpent, there are in fact mysterious wildlife lairs on the bed of the Mekong, giant pools scoured by the enormous whirlpools that form between rapids. There are fifty-eight known pools, and they are immense. Some of them extend for 300 feet below the regular riverbed, making them deeper than the English Channel. They are essential dry-season habitat for huge numbers of fish, including real monsters of the river such as the catfish and the equally rare Mekong population of Irrawaddy dolphins.

But these pools and rapids are threatened by the dynamiting that can now be heard on long stretches of the river in Laos, Burma, and Thailand. China—the river's secretive and overbearing big brother—has embarked on a plan to make the river navigable for ships. That means removing the rapids. Blasting began in 2001 in an area on the border between Thailand and Laos once known for its catfish. Since then, not a single catfish has been seen in that region. And Isabelle Beasley, an Australian researcher studying the Irrawaddy dolphins in the pools, found fifteen dolphins dead in 2003. Ten more bodies were discovered in early 2004. Most of the dead, she says, are caught accidentally in fishing nets. But Beasley's big fear now is the destruction of the pools as the Chinese blasting moves south. "If we lose the pools, we will lose everything," she explains.

And with the dynamiters come the dam engineers. The Mekong and its tributaries have been a mouthwatering prospect for hydroelectric engineers for half a century. For almost 600 miles in China and the mountains of Southeast Asia, the main stem of the river carves a path down narrow gorges, falling more than half a mile in the process. The gorges make the river easy and cheap to dam; the drop means the power that can be generated is stupendous. In the 1960s, the U.S. government's dam builders at the Bureau of Reclamation sought to cement Uncle Sam's tenuous hold on the region by offering to build hydroelectric dams in the canyons. But leaders like North Vietnam's Ho Chi Minh turned the bureau down before war engulfed the region.

Forty years on, Southeast Asia is finally at peace, and demand for water and electricity is soaring. For engineers, the river is the obvious source for both. Thailand, Vietnam, and Laos have already built dams on tributaries of the Mekong, and hydroelectricity is now the biggest legal export from Laos. Most of these dams have proved problematic in one way or another. Vietnam's blocking of the Se San at Yali Falls is causing chaos as unannounced releases of water flood villages over the border in Cambodia. Thailand's Pak Mun Dam on the Mun River wrecked local fisheries. The Nam Theun II, for which Laos wants international funding, has become a cause célèbre before a stone has been laid, because locals fear it too will kill their fish.

China, meanwhile, is building a cascade of eight huge dams on the main stem of the Mekong, which it calls the Lancang. The first two are already operating and generating concern downstream all the way to the sea. As the turbines are switched on and off to meet changes in demand for power and their reservoirs empty and fill, water levels in the river fluctuate by up to 3 feet a day for hundreds of miles downstream. Peter-John Meynell, of the World Conservation Union, who regularly sails on the river, says, "I feel the wash myself going down the river. Even big boats find it difficult because of the surges from the dams. Local fishers are losing their livelihoods as a result."

Meanwhile, the initial filling of the first two dams—the Manwan in 1993 and the Dachaoshan in 2003—coincided with unusually low flows on the Mekong all the way down to the Tonle Sap, and with poor fish catches. In other years, hydrologists have measured a reduction in the Mekong monsoon flow that they suspect is due in part to the dams holding back the floodwaters for release through the dry season. The Mekong's pulse is being weakened.

But this is just the start. Construction on a far bigger dam began in 2002. The Xiaowan Dam will tower 958 feet over the river, as high as the Eiffel Tower. Its reservoir, at 105 miles long, will be twenty times the size of the two existing reservoirs combined and second in China only to the Three Gorges, on the Yangtze. After that will come an 800-foot-high dam, the Nuozhadu, which will have an even bigger reservoir. By early next decade, the cascade of dams will be able to store 32 million acre-feet of water—more than half the river's flow as it leaves China.

According to Chinese operational plans, the river's dry-season flow out of

China will triple after the cascade is finished, and the huge wet-season flow will fall by a quarter. The effect will diminish as you go downstream, but with 40 percent of the river's flow coming from China in the dry season, it will still be enough to halve the strength of the flood pulse in Cambodia as it passes the royal palace. The effect of that on the reversal of the Tonle Sap is harder to predict. In a worst-case scenario, though, the reversal could give out altogether.

Some old-school river scientists argue that these changes could be a good thing, moderating floods and enhancing water supply downstream in the dry season. Ted Chapman, of the Australian National University's Mekong Research Network, says, "My own view is that the Mekong cascade could be very beneficial to the Mekong region." Not so, says Eric Baran, a French biologist investigating the Mekong: "Western engineers think of flood extremes as bad news, and engineers in Cambodia have absorbed this Western philosophy. But in Asia, floods are generally a good thing, because they drive the natural ecosystem on which millions depend for their food." Hans Guttman, a senior hydrologist at the river commission, agrees: "There is a strong relationship between flood flows and fish migration. Flattening the flood peaks would have a severe effect on the river's ecosystem."

And that effect would be felt nowhere more than in and around the Great Lake. Matti Kummu, of the Helsinki University of Technology, who is modeling the river's hydrology, figures that the cascade of dams would not end the Tonle Sap's reversal but would delay it by about a month. That would be enough to reduce the area of flooded forest by more than 2 million acres, he says. But the greatest danger might be the curtailing of the pronounced dry season. Without that, much of the forest would die.

Perhaps just as important as the changing water flow would be the loss of the river's load of silt. And here the Chinese dams are of even greater significance. Around one half of the 160 million tons of silt coming down the river each year begins its journey in China. Much of it ends up in the flooded forest around the Great Lake, fertilizing the vegetation and feeding the growth of fish. According to Kummu, the first two Chinese dams are already capturing half the silt flow from China. They may be as much to blame for declining fisheries as the faltering water flows. "If the whole cascade of dams is built, it would trap some 94 to 98 percent of the sediment load coming from China," he says. That would be a catastrophe for the fertility of the Mekong flood, the flooded forest, and the entire ecological infrastructure on which much of Southeast Asian rural life is built.

———

The Mekong matters to the people who live around it perhaps more than any other river on earth. Some communities in Cambodia are as far as 30 miles from the nearest road. For their inhabitants, the river is the road and the annual flood is the basis of their lives. "Fish are landed virtually every mile along the river by tens of millions of people from tens of thousands of communities," says Barlow. Yet the Mekong fishery's place in Cambodian society and its economy is largely hidden from urban elites and government. Most of the fish caught on the river never reach commercial markets and never appear in government data.

As societies in Southeast Asia urbanize, the lines are being drawn for a final battle for the Mekong. Engineers see the river as a potential powerhouse for industrialization and urbanization, and politicians are taking the lure. Ministers are seduced by "hard figures, megawatts of power," says Barlow. "They are far less impressed with what we tell them about the livelihoods that depend on the fish. They really don't appreciate what is at stake. It's not just about money —it's the food, the protein, and the jobs. The fisheries are a source of natural wealth for the poor. It belongs to them. But if the fisheries are destroyed, their only alternative would be a job in a factory in Phnom Penh making textiles for the West." Full nets or full factories: maybe that's the choice.

IV

When
the rivers
run dry . . .

floods may not be far behind

13

China:
The Hanging River

In early June 1938, at the height of the Sino-Japanese War, Chinese generals perpetrated what remains the single most devastating act of war ever—on their own side. Fighting to hold back Japanese invaders, they sent eight hundred troops to the great dyke that holds in check the Yellow River as it falls onto the wide North China plain. They told the troops to dig pits beneath the four-hundred-year-old dyke and fill them with explosives. And on the morning of June 9, the generals ordered the detonation of the explosives and the firing of artillery shells to widen the breaches. The aim was to destroy the dyke and create a great wall of water to stop the Japanese army.

At first the water flowed only slowly through the half-mile-wide gap in the Huayuankou dyke. It seemed as if the plan had failed. But gradually, as the summer monsoon floods came downstream, more and more of the river escaped the confines of its channel and spread across the plain. By the end of June, villages, towns, and farmland across three provinces had flooded and millions of Chinese had fled their homes.

More than half a century later, I discovered hanging in a small museum near Huayuankou a dog-eared photograph taken in July 1938 from the remains of the dyke. It shows a scene of utter devastation. I have seen only one picture like it, taken shortly after the atomic bomb was dropped on Hiroshima in 1945. The flooding delayed the advancing Japanese for about a month. But thousands upon thousands of Chinese drowned as the Yellow River marauded across the wide plain. And as the flood swelled, famine and disease spread.

Chinese historians put the final death toll at a staggering 890,000 people, virtually all of them Chinese.

Stop a while and take that number in. It is three times the number killed by the Indian Ocean tsunami of December 2004. It is greater even than the combined death tolls of the atomic bombs dropped on Hiroshima and Nagasaki and the firebombing of Dresden and Berlin and Tokyo during the Second World War.

During the floods, the river abandoned its old course altogether and chose a new path, finally reaching the ocean nearly 450 miles farther south than before, near the mouth of the Yangtze River. There it stayed for almost a decade, until 1947, when engineers managed to divert it back within its old dykes. All that remains on the rebuilt Huayuankou dyke today to remind visitors of the destruction once wrought there is a small stone memorial. On the day I visited, most of the people milling around it were uncomprehending young Chinese enjoying the last autumn sunshine. But occasionally older citizens stopped for a while, consumed in thought.

The Chinese have always called the Yellow River their "joy and sorrow." Joy because control of its waters to irrigate crops has sustained more people for longer than anywhere in history. But sorrow because of what happens when the river fails or the control gives way. Unlike the Mekong, where floods are still good news, flooding along the Yellow River is usually disastrous. The river has probably killed more people than any other natural feature on the earth's surface. Controlling those floods has always been the single most important activity of Chinese governments. Many historians argue that it is the single most important reason for the creation and survival over the millennia of the vast Chinese state with its draconian powers. The Chinese sum up the relationship in a word: *zhi,* which means both "to regulate water" and "to rule."

But fast-forward sixty years from the great disaster of 1938 and the news is all about the river running dry. The Yellow River first failed to reach the sea in 1972. Between then and 1998 it stopped short of its delta for part of almost every year. In 1997 it did not reach the sea for more than seven months. For most of that time it trickled into the sand on its bed at Kaifeng, 485 miles inland and just downstream of the site of the 1938 breach.

I set out to discover why the river seems to have changed its character so much—why the killer floods are giving way to drought. But on my journey I began to conclude that the river has changed less than it appears. I found that the two phenomena of flood and drought on the river are inseparable, like yin and yang. And while Chinese commentators rail against the river's running dry, many experts at the Chinese government's Yellow River Conservancy Commission, whose job is to manage the river and prevent disasters, believe that the next great disaster could be another major flood. "The flood threat of the Yellow River is still the hidden danger in China," says the commission's director, Li Guoying.

As the late Harold Dregne, of Texas Tech University, put it in 2001, floods remain "the most fearsome threat on the river. A minor breach of an embankment could cause one of the worst catastrophes in human history—the threat is very real." The truth, I discovered, is that floods and droughts have always gone together here. Every year of low flows increases the chances of a future lethal flood. And modern engineering has only upped the ante.

The Yellow River begins high in the mountains of eastern Tibet. It loops first north through the Gobi Desert and then south and east, through the heart of the world's most populous nation, to the Yellow Sea. It is a journey of more than 3000 miles. The river is the fifth longest in the world. Almost half a billion people depend on it for water to drink and to grow their food. Irrigation water from the river has made China the world's largest producer of wheat and second largest producer of corn. But this success story is threatened today by growing water shortages.

The river is running low all the way from its source to the sea. Up on the eastern edge of the Tibetan plateau, the area where it rises used to be known as "the county of thousands of lakes." But more than half the lakes have disappeared in the past twenty years, and a third of the pastures have turned to desert. In the river's middle reaches, irrigation canals are running dry, fields are being abandoned, and desertification is generating huge dust storms that spread east, choking lungs in Beijing, closing schools in Korea, dusting cars in Japan, and raining onto mountains across the Pacific in western Canada. Near the river's mouth on the fertile expanses of the North China plain, springs no longer gush, the meadows are gone, and few of the old lakes that once dotted the plain survive.

The desiccation of the Yellow River and its basin is an economic as well as an environmental disaster. As its waters have faltered, millions of acres of farmland in the river's lower reaches have been abandoned. The wheat harvest fell by a third in six years from the late 1990s. A new dust bowl looms in the breadbasket of China, reminiscent of the disaster in the U.S. High Plains in the 1930s. The World Bank warns that the desiccation will bring "catastrophic consequences" for China's ability to feed itself. This matters to the world. As China's grain imports soared, the world's granaries emptied and global grain prices rose by a third.

What has gone wrong? For one thing, there has been drought. Historically, the river has an average annual flow of 46 million acre-feet. But in the 1990s, the average was just 35 million acre-feet. The drought may prove to be a cyclical change or something more permanent. It could be an early sign of global warming. But an equal cause of the river's fitful flow is a dramatic rise in abstractions to feed irrigation and China's fast-growing cities. State irrigation projects dotted along the river cover 30,000 square miles and soak up the majority of the river's water—so much so that in dry years, the total official allocations are greater than the actual flow.

China has put a giant bureaucracy, the Yellow River Conservancy Commission, in charge of the river. I traveled the river with its staff, constantly struck by how their comments jumped from civil service reticence to open candor. One of the problems, they said, is that "the most inefficient irrigation schemes are always in the driest places." On the fringes of the Gobi Desert— in the provinces of Gansu, Ninxia, and Inner Mongolia—it typically takes four times as much water to grow a field of wheat as it does in the lower river basin. Yet in many years, all the water is taken in these dry upstream provinces and little or nothing is left for Shandong and Henan on the North China plain, where the soils are better, evaporation is less, and the water would be much more productive.

At the start of my journey, in an office way upstream at Langzhou, the capital of Gansu Province, one commission official pointed to a map of local irrigation projects. One pumped water 2000 feet up from the Yellow River onto sandy hilltops. Another sent the precious liquid through 55 miles of tunnels to reservoirs, where much of it evaporated. Was this a good use of the water? The local officials who built the projects say yes. The water grows crops—end of story. The commission's river managers say no. The irrigation projects are hopelessly uneconomical, built more for prestige than for profit, and are vastly wasteful of water.

Meanwhile, China's fast-growing cities want the same water. Altogether, about 8 million acre-feet of the river's water—getting on for a quarter of the flow today—is piped out of the river basin, much of it to distant cities. And that is putting ever greater stress on the river. Taiyuan and Hohhot, the capitals of Shanxi and Inner Mongolia, are both digging canals 60 miles long to connect to the sputtering lifeline. Taiyuan, ironically, needs the water only because the city's old source, the Fenhe River, a tributary of the Yellow River, has dried up as a result of huge abstractions for the state's vast coal mines.

To see some of the downstream consequences of this upstream mismanagement, I went to the People's Victory Irrigation Project. It is just over the river from the Huayuankou dyke, where the river flows out onto the North China plain. The project is a fifty-year-old Communist totem—one of Chairman Mao's first public works to feed his people. Its canals irrigate an area about the size of Greater London, where a million people live and work. But things are in a sad state. The main distribution canal is badly polluted with foam from a local paper factory. Much of the equipment has not been replaced since the fifties. The original iron sluice gate that the "great helmsman" Mao opened more than half a century ago is still in place. And, worse, the water is giving out.

In the old days, the project was literally awash with water. It sucked almost 800,000 acre-feet from the river each year. The entire project was shut down for a while because it became waterlogged. No such luck today. "We are always short of water now," director Wang Lizheng said with a sigh, in between nervous puffs on his People's Victory cigarette. "We are allowed to take only half as much water from the river as we used to." Shortages are made worse because more than half the water disappears into the ground through the bottom of leaky canals. Wang's main task today is finding enough cash to enable him to line the canals. So far he has lined only a fifth of the main canal. "We do about a mile more a year," he told me. At that rate, it will take another twenty years to finish the job. Farmers are drilling wells to catch the water that seeps from the canals, but even so, only about a quarter of the original area is still farmed.

In Tianzhuang village I met Shang Sumei, who was carrying a small child on her hip. "Most people now work in factories and only farm in their spare time," she said. She is a teacher, and her husband works in a paper factory. "My children won't be farmers. They will go to college and work in the city," she said. Her neighbor Jiao Zhilong said there were often water shortages, so he spent most of the time selling electrical goods in the local market. Another, Pu Yujun, dismissed his agricultural activities in a sentence and instead took me on his motorbike to see his workshop, where eight people mill wheat to make noodles. In the yard outside, I met a man who had his own workshop making plastic sheeting for electrical cables. "There are five hundred households in this village and fifty factories," he told me. Anything but farming.

Mao's greatest investment to feed his people, his symbol of what he saw as the "people's victory" over nature, is now quietly decaying. Mao is still revered here, but later leaders, who starved his favorite project of both funds and water, are not. Wang took me to the tiny project museum, where he showed me a chair—roped off beside old black-and-white pictures of happy peasants digging dykes with spades and bullock carts carrying bumper harvests—on which Mao sat after he opened the sluice gate in 1952. It has pride of place still (even though, as one of the party gleefully pointed out, when one of Mao's successors later sat on the chair, "he broke it with his fat ass").

Symbolism counts for a lot in China. As we drove on, my guide drew the lesson: "The government is becoming weak. It cannot maintain these projects anymore." An old dictum holds that governments in China live or die by their control of the waters of the Yellow River. If so, then here, at least, the current regime is in big trouble.

———

Downstream from the People's Victory project, on the North China plain, water shortages are even greater. In a good year, the coastal province of Shandong can produce a fifth of China's corn. But in the 1997 growing season, the river failed to reach the province, which is the country's largest, with a population of 90 million. Crops died in the fields. There was panic in the government. An edict was issued from Beijing: The Yellow River shall not dry up. Ever. The government for the first time began to enforce limits on how much water each province could take from the river, and it told the Conservancy Commission to manage the river so that a minimum flow of 13,000 gallons a second always reached the sea.

In the headquarters of the commission in Zhengzhou, I watched the giant control system in action. One wall is covered with a huge electronic map of the river, with real-time readings of key hydrological data. An alarm sounds if any of the dozens of automatic monitors on the lower river detect that the flow is getting down to the limit. The operators know who is taking water from the river and how much at any moment and can press a button to shut sluice gates and reverse pumps to stop all abstractions. To make the point, the engineers reset the emergency shutdown limit to just above the flow on the day of my visit, so I could hear the alarm sound. The sense of power over a mighty river was palpable.

The government can now truthfully say that the river never runs dry. Since 1999, it has always flowed. But despite the macho politics and state-of-the-art engineering, not much has changed. The river's managers say they still don't have effective control of what happens far upstream, where most of the water is wasted. Their political writ often does not run up there. And the tiny trickle that always gets through to the delta is "largely symbolic," they admit. Certainly it hasn't stopped the sense of crisis.

As the river empties, farms and cities alike try to keep going by pumping water from beneath the ground. In Shandong on the delta, more than half the irrigation water now comes from underground. But the inhabitants are pumping up water twice as fast as it is being replenished by rains and the river. The underground reserves on the North China plain are being emptied 24 million acre-feet a year faster than the rains replenish them. In the 1960s, the water table was almost at the surface; now it is 100 feet down. In places around Beijing, 90 percent of the replenishable water is gone, and here and there the city is tapping water half a mile down in fossil aquifers that will never refill. Fearing the worst, the city has banned new water-guzzling industries and even emptied the lakes in front of the Summer Palace.

In the summer of 2000, the year after the minimum flow was established, water wars broke out in Shandong as thousands of farmers made illegal connections to a reservoir that had been earmarked to supply cities. When police intervened, there were riots. One police officer died, and more than a hundred protesters were injured. Meanwhile the problems are spreading upstream, where low flows are becoming ever more frequent. In both 2002 and 2004, the river nearly dried up in Inner Mongolia—a thousand miles from the sea. China, superficially the most ordered of nations, seems to be subsiding into hydrological chaos.

———

They don't call the river Yellow for nothing. Its waters are thick and silty. Standing at the Hukou Falls, in the river's middle reaches, a popular spot for Chinese tourists, I watched what looked like chocolate mousse churning over the giant boulders into a narrow gorge. Every year roughly a million tons of silt flow over these falls and on down the Yellow River. It is by some degree the world's muddiest river. Every ton of water contains about 90 pounds of silt, twice the load of its nearest rival, the Colorado River, and seventy times that of the Mississippi. The silt, like the water, brings joy and sorrow: joy because it makes the lower valley supremely fertile, but sorrow because of what it can do to the river when it is not properly managed. As the river's successful rulers have always known, managing the silt is as important as managing the water.

My first destination on a journey that uncovered the truth of this was a region near the middle of the river, where it cuts through steep terrain known as the Loess Plateau. The plateau is the source of 90 percent of the silt in the world's siltiest river. Nowhere on earth loses as much soil to erosion. This is because the Loess Plateau is not a proper mountain range at all. There is no underlying geology. It is just a huge pile of loose sand, several hundred yards thick and covering an area five times the size of Louisiana.

The sand blew here from a distant desert thousands of years ago and has been left out in the rain ever since. Four fifths of it is bare, without any kind of vegetation. Every drop of rain falling on the barren slopes eats away at the sand pile. During the monsoon season, the water washes in torrents down the plateau's quarter of a million gullies (that's the total so far, but only those more than a third of a mile long have yet been counted). These muddy torrents can carry almost as much silt as water.

Erosion on this scale changes the landscape fast. Five thousand years ago, this pile of sand was a true plateau, dotted with fortresses that have since earned it the title "the cradle of Chinese civilization." But today the "plateau" has become a maze of steep-sided hills and valleys with occasional surviving flat hilltops, some still occupied by villages and fields. For China, the erosion of the Loess Plateau is an environmental disaster—both for the plateau itself and for the river downstream, where the silt clogs reservoirs and the riverbed itself.

The silt in the Yellow River has been a constant source of embarrassment to China's leaders ever since they ordered that dams be built on the river. The faces were reddest after the completion of the Sanmenxia Dam. The first dam on the main stem of the river, it was intended as another engineering triumph for Chairman Mao—simultaneously a means of regulating water for the People's Victory irrigation district downstream, for preventing floods, and for generating hydroelectricity. Mao expelled 400,000 people to make way for the reservoir, and was so proud of it that he had its image printed on the country's banknotes. It was part of his 1950s campaign for "the mountains to bend their tops and the rivers to give way."

But this particular river did not give way. Within two years it had filled the Sanmenxia reservoir to the brim with mud, which eventually backed up for 60 miles, causing flooding on the Wei River, a tributary of the Yellow. Chinese engineers have since learned how to avoid catching the worst of the river's silt flows in reservoirs, but even so, the twenty large reservoirs on the river currently hold more than 11 billion tons of the stuff. Since the Sanmenxia disaster, China's leaders have dreamed of stabilizing the Loess Plateau to eliminate erosion and make the Yellow River run clear.

To this end, they have spent half a century remaking the steep eroding hillsides into staircases of terraces, planting swaths of trees and bushes, and plugging the gullies with small dams known as check dams. This work began back in the 1960s with another of Mao's slogans: "Let bald hills become green fields." In most of the world—from Peru and India to the Philippines and the Holy Land—ancient agricultural terraces are being abandoned. But here in the Loess Plateau they are still being built, probably in record numbers.

It is a huge, continuing enterprise of the kind that only a Communist society could attempt. Upwards of 30,000 workers are involved at any one time, often digging terraces and forming chain gangs to plant trees on the steep slopes. So far a third of the fields have been terraced. And the plan is to have 60,000 check dams to trap soil flowing down gullies by 2010.

The World Bank has funded much of the recent work. James Wolfensohn, the bank's president, lauded progress during a visit to China in 2004. A decade before, he said, "The hills in an area the size of Switzerland had looked arid, terrible, stark with no trees. Now they've got grass and trees and animals. It's an area where you'd like to have a holiday... it looks like Switzerland." I went to see.

I began on a hilltop in Dingxi County in Gansu with the county's head of soil and water, Cheng Hangjun. In the valley below, dozens of old terraces had been renovated over the previous year and planted with cypress trees. "Until last year people grew crops right up here. Now that's banned," Cheng said. The aim was to trap 80 percent of the soil and water that used to flow off the hills in the heavy rains. "You can see an effect already. Only clear water flows into the valley now."

The official word is that peasant farmers are happy to have their hillside plots planted with trees, and certainly in areas where I was taken in Gansu and neighboring Shaanxi, that seemed to be true. They were being compensated with deliveries of as much grain as they would have grown on the plots they had given up, and they got help in growing high-value produce that they could sell in nearby towns.

Farmers said that the terraces had made them richer, too. Liu Kequang, in Kongge Lao village in northern Shaanxi, told me that "the terraces trap water, so we get better yields on the fields where we still grow crops. I used to get 450 pounds of corn for every mu [a sixth of an acre], but now I get more than 900 pounds." He also grew potatoes and apples, which he sold to a private trader who took them to southern China. Liu's diminutive eighty-five-year-old grandmother wandered around the homestead tending pot plants, a reminder of the days before Mao and the Party. But his brother had just bought the family's first car.

A few valleys away, in the village of Ermaoqu, I met Wang Shiping. His village had recently replanted grain fields with cash crops such as green beans, peanuts, and watermelons and turned many of the old hillside terraces over to apple orchards. Three times a week Wang went to the nearby town of Qinghuabian to sell his melons. "Five years ago we had only ragged clothes, but now we buy good clothes in the town market. We can all afford TVs and motorcycles. My income has increased four times," he said while wielding a massive chopper to cut melons in half and distribute them to our party. The village showed the sign of wealth that holds across most of rural China: it had more motorcycles than donkeys.

For several days local officials took me to the tops of hills to show off their terrace-digging and tree-planting prowess and to introduce me to happy farmers. The work is genuinely remarkable, and the gains for some farmers, at least, seem unquestionable. But the declared purpose of all this work is to make the Yellow River run clear all the way to the sea. So does it? Can it? To find out, I had to return to the river's lower reaches—and to go back in history: the history of why silt matters so much.

———

The story of China's love-hate relationship with the Yellow River and its silt goes back eight thousand years to the dawn of farming on the river's floodplain. The first farmers set up here to take advantage of the smear of silt brought down from the eroding Loess Plateau and dropped onto the floodplain as the river meandered back and forth. But while the farmers grew good crops in the fertile silt, they were at the mercy of a river that flooded huge areas during the annual monsoon and regularly changed course. So they started to build dykes to keep the river on a single path.

But from that point on, the river stopped spreading its silt. It left it instead on the bottom of its new permanent channel. The layers built up. Every year the riverbed was a little higher. It soon began to rise above the level of the surrounding floodplain, and the "hanging river" was created. It was a point of no return. From then on, the dykes had to be raised higher to keep the river in. The task of holding back the river could never be shirked.

The consequences of failing to contain the river became greater and greater as time passed and the river rose higher. There were repeated disasters. Between 600 B.C. and A.D. 1949, the river changed its course across the North China plain twenty-six times—roughly once a century. The regular cataclysms often brought down dynasties. In ancient times, if the river shifted ground, the emperor was thought to have lost the mandate of heaven and could no longer rule. And sometimes there were global consequences. The

Black Death, which killed up to half of Europe's population in the fourteenth century, may have begun when rats that carried plague fled China after the Yellow River changed course.

As recently as 1855, the river broke its dykes so decisively that it abandoned one route to the sea almost overnight and took over another channel that entered the sea 600 miles farther north. Since then the river has resumed its ascent above its floodplain, held back only by dykes that have to be constantly raised and maintained by press-ganged peasant labor. In places the river is now some 65 feet above the land outside the dykes. It has never been as high as it is today; nor has it ever risen as fast as it has in the past couple of decades. Never have the human and political consequences of a breach been so great.

Why the apparent rush to disaster? The problem is the rate at which the modern Yellow River is accumulating silt in its lower reaches. Until the mid-twentieth century, at least half of the silt flowing down the river made it to the sea. But today only about 10 percent gets there. Most of the rest accumulates either behind dams or in the hanging river. This accelerated buildup of silt is happening, say scientists at the Conservancy Commission, because less water is coming down the river, and because when it comes, it is traveling more slowly. In the 1950s it took floods seven days to make it to the sea; now it takes eighteen days. A slower river has less force to bear the silt to the sea.

With 90 percent of the sediment now ending up on the riverbed, the hanging river is rising rapidly. By early this decade it was rising by almost 4 inches a year, or over 3 feet every decade. Dykes can no longer be raised to keep up, so the capacity of the river channel is getting smaller. Its ability to contain summer floodwaters is falling alarmingly. In 1950 the Yellow River flow could reach 6.5 acre-feet a second before it spilled over the inner dykes. By 1990 that was down to 3.2 acre-feet, and by the end of the 1990s to a measly 1.6 acre-feet.

In recent years drought has kept river flows low enough to avoid disaster. River engineers have helped by emptying their reservoirs before each monsoon season, so they are ready to catch whatever floods come down. But this only exacerbates the problem in the long term, because the slow flow adds to the silt accumulation.

A crisis is approaching. A breakdown of the dykes and a breakout of the river seem imminent. It was something I pondered as I stood on the Huayuankou dyke. On one side the water was surging past. On the other, 87 million people were living in a flood zone as large as Louisiana. I looked at the marks on the dyke showing high water levels during past floods in 1958 and 1996. The 1996 flood contained only a third as much water as the 1958 flood, but it rose 3 feet higher up the dyke wall. In September 2003, just a year before my visit, a flood of just 2 acre-feet a second broke an illegal dyke erected by villagers and inundated 100,000 people near Caiji village in nearby Henan. Ten thousand soldiers fought the breach and prevented disaster.

———

What should be done? Some believe it is time for another change of channel —preferably a managed change rather than one born of catastrophe. Richard Cowen, at the University of California at Davis, argues that China should abandon the hanging river and start digging an entirely new route for the river to the sea. But, he says, "It seems far more likely to me that there will have to be a catastrophe before China summons the resources for a major renewal of the river."

In China there are two approaches. One is to cut the amount of silt coming down the river and clogging the channel. That is the national purpose of the great project on the Loess Plateau. But is it working? A lot of inflated claims are being made for public consumption, and the official line from the Conservancy Commission is that the work on the Loess Plateau has cut silt flows downstream by a fifth, to 1.3 billion tons a year. But when I asked about this within the commission, its scientists were decidedly cautious.

Xiong Weixin, the deputy director of its bureau of soil and water conservation, told me that average figures mean nothing. Annual silt discharges range from 3 billion tons in a wet year to less than half a billion tons in a dry year. The low figures are mostly from recent years. That could reflect the conservation work, but is was at least as likely to be the result of poor rains. Most of the silt comes downstream after a few intense storms, and many farmers told me during my travels that they hadn't had a big storm for five years. Nobody can yet disentangle the two effects.

Even optimists at the commission doubt that it will be possible to cut silt flows much further. Massive erosion, they point out, is a natural feature of the Loess Plateau, and to try to halt it would be decidedly unnatural. The dream of making the river run clear, they say, is just that: a dream. The Chinese, brought up on the wisdom of managing the Yellow River, sensibly have an idiom, "when the river runs clear." It means "never."

The second approach is to increase the amount of water coming down the river. More water will flush the silt to the sea. "It's like an inoculation," one commission scientist said. "We have to have regular small floods in the river in order to prevent major flood disasters." The problem here is that the river does not contain enough water these days for proper flushing. The current estimate is that efficient flushing requires at least 15 million acre-feet of water to reach the sea each year. That's more than a third of all the water that enters the river in a typical year, and almost half of the river's flow in some recent years. As the Conservancy Commission puts it, "This challenge will be vitally serious and difficult to meet." For that, read politically impossible.

So the river's engineers are trying to use the dams to send an artificial flood pulse to flush the river of silt. It's a genuinely innovative idea, never tried on this scale anywhere in the world. They store up water behind the dams and then release it all at once—like a giant cistern on a toilet—to flush away the sediment on the riverbed downstream. At the start of the monsoon seasons of 2002, 2003, and 2004, they successfully flushed away some 220 million tons of silt, lowered the bed by a foot in places, and increased the capacity of the channel to some 2.4 acre-feet a second.

The Conservancy Commission engineers are divided about how much further the flushing can go. But they hope that annual flushing from the reservoirs can stabilize the channel's capacity at its current level. That would be a help, but the capacity of the river channel to handle floods is still not much more than a third of what it was half a century ago. While the droughts continue and the upstream reservoirs can handle what floods do come downstream, that may not matter much. However, it means that the river is primed for disaster if a big flood ever comes.

So could an enormous human disaster happen again? The answer has to be yes. And it is the years of low flows that have made it possible.

I asked the engineers what floods they feared most. What might trigger a disaster? The river, they say, has two Achilles heels. The first is its vulnerability to ice. On its long journey to the sea, the river twice makes giant excursions to the north, first into Inner Mongolia, in its middle reaches, and second as it crosses the North China plain to the sea. Sometimes the water in these northern areas freezes and blocks the river. With water continuing to flow downstream, the result is floods. Historically, about a third of the big floods on the Yellow River have been a result of ice blockages. They are notoriously difficult to prevent—so much so that there was a saying in the Qing Dynasty that a river official who failed to anticipate a flood, who would normally lose his life as a consequence, could not be executed for an ice flood. In modern times, there were big ice floods in 1951 and 1955. As recently as December 2002, a buildup of ice along 60 miles of the river in Shandong threatened to cause flooding. Half a million people were mobilized on Christmas Day to fight threatened floods by strengthening dykes and dynamiting ice. It worked.

The second Achilles heel is intense local flooding that no dam can contain. Li Guoying, the director of the Conservancy Commission, said in a recent paper on the long-term security of the river that the flood defenses have a weak point between the lowest dam, the Xiaolangdi, and the Huayuankou dyke —the one the generals chose to break back in 1938. "Although this area is not too big, it is a district with extreme rainstorms," he said. One rainstorm not far away in June 2000 produced 20 inches of rainfall in twenty-four hours and dumped about 5 million acre-feet of water into local rivers, luckily without disaster. But, said Li, "If that storm had moved 60 miles west, to the area between Xiaolangdi and Huayuankou, [the consequences] would have been unimaginable."

———

Chinese society along the Yellow River is the supreme example of what historians call a "hydraulic civilization," a centralized autocratic society built on the paramount need to control its water resources, to harness them for food and tame them to prevent destruction by floods. The struggle is encapsulated by the legend of Yu the Great, a leader credited with controlling the floods on the Yellow River four thousand years ago by dredging channels and allowing the water to flow to the sea. On a ridge near Zhengzhou, not far from the headquarters of the modern Conservancy Commission, his statue towers above the floodplain. One hand is outstretched, bestowing prosperity on his people. But the other hand carries the tools with which he cleared the river. His face looks defiantly toward the river. Taming it had been a grim, remorseless task. And once undertaken, it is a task without end.

Standing on the dykes outside Zhengzhou, I considered Yu's legacy and imagined the huge endeavor by hundreds of millions of people over thousands of years to keep the Yellow River in check. The dykes constructed in the past fifty years alone have a volume equivalent to thirteen Great Walls of China. But the risks that China runs with this river seem to be increasing. I have no doubt that the engineers do their best, but the country seems to be sleepwalking to the next disaster, believing against all the evidence that history will not repeat itself. Dyke and dam building cannot save China forever from the wrath of the Yellow River. And it is hard to escape the conclusion that the next disaster may not be so far away.

14

Changing Climate

Nobody can say for sure how much of the declining rainfall across the basin of the Yellow River is due to manmade climate change and how much to natural fluctuations. In the short term, it is probably unknowable. What can be said is that a good number of rivers have experienced serious declines in the amount of moisture reaching them in recent years, and that much of this is in line with the predictions of climate researchers.

Equally clearly, rivers in these desiccated regions are also being emptied by direct human activity, often to irrigate fields parched by the lack of rain. We have looked at how the combined effects have reduced flows on the Rio Grande in the American West, the rivers feeding Lake Chad in the African Sahel, and the rivers of Central Asia, such as the Helmand in Afghanistan.

But as climate change gathers pace in the coming decades, desiccation won't be universal. Modelers in organizations such as Britain's Met Office predict that in some cases there will be increased river flows. The one certainty is that change is taking place quite rapidly on many rivers, and the hydrological statistics on which dam builders, river managers, and irrigation districts rely to build their structures and make their plans often have little meaning anymore.

First the global picture. Higher air temperatures will increase evaporation from the world's oceans and so intensify the water cycle. By later this century, there could well be 8 to 10 percent more water vapor in the atmosphere on an average day than there is today. That is 800 million acre-feet or so extra, enough to fill twenty Niles. This will almost certainly increase global rainfall. Also, shifting trajectories of rain-giving climate systems, like Atlantic cyclones, will mean that the total rainfall will be redistributed. Many middle latitudes will become drier. Meanwhile, the higher temperatures will also mean faster evaporation of water on land, so soils will dry out more quickly. That means less of the rainfall will reach rivers.

Future flows on individual rivers will depend in large part on the balance between the competing effects of changing precipitation patterns and greater evaporation. The rule of thumb seems to be that dry areas will become drier while wet areas will become wetter. Globally, climate will become more extreme, and rivers will respond in kind.

As a result, many of the rivers that provide water in the world's most densely populated areas and where water is already in the shortest supply will be in still deeper trouble soon. In northeastern China, the savanna grasslands of Africa, the Mediterranean, and the southern and western coasts of Australia, rains will probably diminish, evaporation will certainly be greater—and the rivers will run dry.

The U.S. government's Scripps Institution of Oceanography estimates that reservoir levels in the Colorado will fall by a third as declining rainfall and rising evaporation combine to reduce moisture by up to 40 percent across the southern and western states. The Niger River, which waters five poor and arid countries in West Africa, is expected to lose a third of its water, and the Nile, the lifeblood of Egypt and Sudan, could lose a fifth. Inland seas will be at special risk as many rivers in continental interiors lose flow. Lake Chad has almost succumbed and may suffer further in future. Also under threat are the Caspian Sea and Lake Balkhash in Central Asia, Lakes Tanganyika and Malawi in East Africa, and Europe's largest, Lake Balaton, in Hungary.

In contrast, giant tropical rivers like the Amazon and the Orinoco in South America and the Congo in Africa will become even more bloated than today, the models predict. Similarly, the great Arctic rivers of northern Canada and Siberia will probably gain water as warmer air holds more moisture and more rain falls on their catchments. So the Mackenzie and the Yukon in Canada and the Ob, the Yenisei, and the Lena in Siberia will rage even more fiercely—40 percent more, according to one big study.

There is some evidence that these trends are already well under way. In early 2005, the U.S. government climatologist Kevin Trenberth showed that after a century of little change, there has been a surge in instances of severe drought around the world since 1970. The proportion of the earth's land surface suffering very dry conditions rose from 15 percent then to 30 percent at the start of the twenty-first century, he said. Meanwhile, scientists from the Met Office reported that "far northern rivers are discharging increasing amounts of freshwater into the Arctic Ocean due to intensified precipitation caused by global warming." There has been an annual increase in river flow into the Arctic of 7.3 million acre-feet since the 1960s, they found.

But there will certainly be anomalies. In some places, neighboring rivers could have very different futures. The Indus and the Ganges both rise in the Himalayas and flow south. But while projections indicate that the Indus could be flowing at half its present levels by the end of the century, the Ganges could gain half. In Europe, the rivers of the north and west will probably run fuller as more intense Atlantic cyclones increase winter rains, while those in the south and east will waste away as summer evaporation goes into overdrive.

And some rivers will live a switchback life in a greenhouse world. They will become increasingly unpredictable and dangerous, with flood and drought alternating even more than they do today. That will sometimes be good for wildlife. It could even help revive flood pulses lost to dam building. But the danger is that the pulse will prove too irregular and unpredictable for either nature or people to benefit.

Some rivers will change dramatically with the decades, first running full to overflowing and then shriveling. This is most likely on those that drain the great glaciers of the Himalayas, Tibet, the Alps, and the Andes. As their glaciers melt in the early decades of this century, the flow of some are predicted to increase, especially in summer. Already the Alps have lost a quarter of their ice. Alpine melting in 2002, during the warmest-ever summer in the Northern Hemisphere, contributed to unprecedented floods in central Europe. But as the glaciers disappear, the meltwaters will disappear, too.

A British study in the Himalayas found that the Indus could increase its flow by between 15 and 90 percent in the first half of this century as glaciers melt, and then decrease to between 30 and 90 percent of the current flow later in the century. Likewise, American researchers believe that the spring meltwaters from the Sierra Nevada snowpack, which today sustain summer irrigation of crops and lawns across the desert lowlands of California, could diminish by 70 to 80 percent over the next fifty years. As the rushing meltwaters become a trickle, some of the most productive agricultural lands in the world could dry up.

One important effect of existing glaciers is to stabilize river flow between years. They absorb highly variable monsoon rains and provide a strong, regular flood pulse in the summer melting season. This is vital to people living downstream. The glaciers of the Himalayas and Tibet feed seven of the greatest rivers in Asia—the Ganges, Indus, Brahmaputra, Salween, Irrawaddy, Mekong, and Yangtze—ensuring reliable water supplies for 2 billion people. But in half a century or so, the glacier flows in many of these rivers will dwindle and be replaced by much more fickle flows from rain in the mountains. That is a serious threat to Asia's future.

"Once the glaciers go, you're down to whatever happens to fall out of the sky," says the British glaciologist Martin Price. And not just in Asia. Down the Andes, cities like La Paz, Lima, and Quito—the capitals of Bolivia, Peru, and Ecuador, respectively—depend on glaciers for secure water and hydroelectric power, but the glaciers are disappearing fast.

Uncertainty, then, is the greatest difficulty. Some of these predictions assuredly will not come true. Equally certainly, there will be real-life horror stories unconsidered in the climate models. This creates huge dilemmas for water engineers. How can they design dams and irrigation schemes, with expected lifetimes of fifty or a hundred years, to cope with such uncertainty? In all probability, they can't.

———

Already the white elephants, the victims of recent climate change, litter the landscape. In Ghana, in West Africa, the Akosombo Dam has been left high and dry by declining flows down the Volta River. The dam was designed by British colonial administrators who wanted to kick-start industrialization in Ghana as it became independent in the 1960s. Instead, it virtually bankrupted the nation. The dam was built at a time when rainfall was at a historic high. But these days the reservoir is rarely more than half full. Its electricity output is a fraction of that anticipated. Industrialization never happened.

Likewise the Victoria Dam, which was built by the British in Sri Lanka in the 1970s. It too was a bid to industrialize a newly independent nation. It too languishes half used most of the time, because the Mahaweli River has only 40 percent of the flow anticipated by British hydrologists. And near Lake Chad, as we have seen, the British-designed South Chad Irrigation Project sits high and dry because the lake has long since receded beyond the reach of its intake pipe. These boondoggles warn of a growing crisis of confidence about cherished notions of how much water rivers can be expected to deliver.

Most of the world's population currently lives where there is a history of guaranteed water. That is not by chance. Humans require reliable and predictable water in order to flourish. Modern, highly engineered methods of exploiting water often test reliability to the limit. Now climate change is undermining the predictability, too.

V

When
the rivers
run dry...

engineers pour concrete

15

Wonders of the World

He cut a startling figure. Daniel Beard, the recently retired commissioner of the Bureau of Reclamation, the U.S. government agency that has built more large dams on more rivers than any other body in the world, had taken off his suit and donned a T-shirt and was striding through the streets of Nagaragawa in southern Japan at the head of a thousand protesters. He was yelling, "No more dams!"

The demonstration was headed for a newly completed concrete barrage on the estuary of the Nagara River. It was a project with an unusual history. Costing $1.7 billion, the barrage began as a scheme to supply water to the city. When it transpired that the water was not needed, it turned into a flood protection project. The saga reeked of Japan's seemingly insatiable desire to keep the construction industry hammering away, almost regardless of whether its structures were needed.

Beard was in no doubt that the barrage should never have been built. "This is one of the most awful dam projects I have ever seen—and I've seen some," he told the demonstrators that afternoon through a crackling megaphone. "It will supply water that nobody needs, will probably destroy a salmon fishery on a beautiful river, and may actually increase the risk of flooding."

Beard's public conversion to the antidam cause had begun a few years before, in 1995. He was at the time still in charge at BuRec and was attending a meeting in South Africa of the International Commission on Large Dams, a professional association of the global high priests of dam engineering. One quiet morning he declared to an astonished workshop audience that "the dam-building era in the United States is over." Such heresy. The agency that had raised the Hoover Dam across the Colorado and built the Grand Coulee on the mighty Columbia, that was the United States' largest water wholesaler and fifth largest electricity generator, and that had poured concrete and rock and earth across thousands of other rivers in the States and around the world had, he said, stopped building forever. No more dams.

In Japan, a country with dams on virtually every river, Beard went further, condemning a huge global industry in which his organization had played a central role. He linked arms with Japanese opponents of the Nagara barrage; with Phil Williams, then the president of the International Rivers Network and an archfoe of the old BuRec, who believes every dam is a bad dam; and with Dai Qing, a former missile engineer from Communist China who became that country's best known and most courageous opponent of dams. Beard was, he said, proud to stand that day beside the antidam protesters of the world.

⁓

Modern water engineering began in earnest with the Hoover Dam—the first superdam. The 700-foot plug across the Boulder Canyon on the Colorado River was completed in 1935. It was taller than a sixty-story building, bigger than the Great Pyramid of Egypt, and contained enough concrete to pave a highway from San Francisco to New York. Behind it grew Lake Mead, which could hold more than twice the river's annual flow. As Francis Crowe, the BuRec surveyor on the project, later put it, "I was wild to build this dam—the biggest dam built by anyone, anywhere." This was not the language of technical reports, but it was perhaps the truest reflection of the motivation behind much large dam construction on the Colorado and around the world.

The white concrete of the Hoover Dam became a symbol of Franklin Roosevelt's New Deal public works projects, and of a broader lust to remake the landscape. It was soon joined by the equally talismanic Grand Coulee Dam on the Columbia River. Famed folksinger Woody Guthrie even wrote a collection of songs, the Columbia River Songs, for the Columbia River and the Grand Coulee Dam; in it he called them the newest wonders of the world.

These dams ushered in a postwar world in which dams became symbols of modernism, of economic development, and of mankind's control over nature. Russia wanted its own Hoover Dam; so did Egypt and Japan and China and India. Hoovers sprouted across Latin America; Britain built replicas for its colonies as parting gifts before independence. No nation-state, it seemed, was complete without one.

"Except for global warming, there has been no more drastic human alteration of the landscape in the last fifty years than the damming, regulation, and diversion of the world's rivers," Beard told the Japanese demonstrators. But he was not boasting; he was apologizing. "Those who promoted dam projects were not honest about costs and benefits. The truth is that dam proponents would say just about anything to get a dam project approved."

The decisions to build were "political, benefiting particular politicians or their benefactors rather than solving a problem. In our experience at BuRec, the actual total costs of completing projects exceeded the original estimate typically by 50 percent. And the actual contribution made to the national economy by these dam projects was small in comparison to the alternative uses that could have been made with the public funds they swallowed up. We are now spending billions of dollars to correct the unanticipated impacts such as lost fisheries, salinized soils, and desiccated wetlands." Ouch.

Rivers today are worth more in America as amenities for fishing and tourism than as water for filling reservoirs, Beard said: "We've started tearing down dams." Is the United States a special case? Far from it. "In my view, we are starting down a similar path throughout the world," he added. "The time when large dam projects are a realistic answer to solving water problems is behind us."

Like Beard, modern environmentalists have come to see large dams as engines of environmental destruction. But it was not always so. For many years, environmentalists in both North America and Europe supported the dam builders. Rather like wind power today, hydroelectricity was seen as a new, clean, cheap source of electricity. In Europe, concern that dams might interfere with the natural flow of rivers such as the Danube, the Rhine, and the Rhone was tempered by the fear that the alternative was pollution from fossil-fuel power stations. As late as 1996, the head of WWF (the World Wildlife

Fund) in Austria told me, "Even though we love the mountains, most environmentalists in Austria still support the construction of dams in their valleys."

Hydroelectric dams do generate a huge amount of electricity—around a fifth of the global total. More than sixty countries depend on them for more than half their power. One dam, at Itaipu on the Parana River between Brazil and Paraguay, has a generating capacity of 12,600 megawatts—equivalent to a dozen conventional power stations. It supplies São Paulo and Rio de Janeiro, two of the world's megacities. China's Three Gorges on the Yangtze will be even bigger. Most modern dams serve double or triple functions, also supplying water for irrigation projects and city faucets and sometimes claiming flood prevention capabilities as well.

The world's large dams now hold more than 5.5 billion acre-feet of water. Most big river systems, including the twenty largest and the eight with the most biological diversity—the Amazon, Orinoco, Ganges, Brahmaputra, Zambezi, Amut, Yenisei, and Indus—all now have dams on them. Most of the surviving untamed rivers of the world are in the empty Arctic tundra and northern boreal forests. The largest is the Yukon, in northern Canada. The handful in tropical lands are rapidly becoming extinct. The Salween, which runs for some 1600 miles from China through Burma and Thailand, is about to be dammed for the first time. On rivers like the Colorado, the Volta in West Africa, and the Nile, the big dams can hold two or three times the actual annual flow. And yet they remain an essentially experimental technology. Their hydrological, ecological, and social effects have been huge. But for many years their status as symbols of modernism insulated them from serious appraisal. Even in the 1990s, fewer than half of all proposed dams had an environmental impact appraisal before construction began. And even fewer had the consent of the people they displaced. Only since the late 1990s have serious steps been taken internationally to establish whether their benefits outweigh the environmental, social, and economic costs.

The focal point for this reevaluation has been the World Bank. In the second half of the twentieth century, the bank spent an estimated $75 billion on building large dams in ninety-two countries. But by the late 1980s, its own cost-benefit analyses questioned the value for money represented by this investment bonanza. They catalogued huge cost overruns, billion-dollar corruption scandals, poor design and bogus hydrology that left reservoirs empty, turbines that were never connected to national grids, and promised downstream irrigation projects that never got built at all. Bank-financed dams, the analysts also discovered, had caused the forced resettlement of 10 million people.

Amid a rising chorus of opposition around the world, the bank pulled out of contributing to the funding of a high-profile dam on the Narmada River in India. In a quandary over how to proceed, it appointed a World Commission on Dams to assess the successes and failures of large dams and come up with some ground rules for what a successful dam project might look like.

The final report from the commission was launched in a blaze of publicity in London in late 2000 under the benign gaze of Nelson Mandela and the rather sterner visage of the World Bank's president, James Wolfensohn. It was even more scathing about large dams than the bank could have feared at the outset. It endorsed many of the environmentalists' most trenchant criticisms. Most dams just don't deliver as advertised, the commission said. Average cost overruns were 56 percent. Half of hydroelectric dams produced significantly less power than promised; two thirds of those built to supply water to cities delivered less water than promised; a quarter of them delivered less than half what their brochures claimed. Dams built to irrigate fields were no better. A quarter of them irrigated less than 35 percent of the land intended. Even dams that promised to protect against floods "have increased the vulnerability of river communities to floods," often because their reservoirs have been kept full to maximize hydroelectric production.

And dams have taken huge amounts of land—land on which people once lived. All told, at least 80 million rural people worldwide had lost their homes, land, and livelihoods, the commission found. The Akosombo on the Volta in Ghana expelled 80,000 people, the Aswan High in Egypt 120,000, the Damadur in India 90,000, the Kariba in southern Africa almost 60,000, the Tarbela in Pakistan more than 90,000. And for what? My own estimates show that the Aswan High Dam generates just 2 kilowatts of electricity for every acre of land flooded; Kariba generates just 1.2 kilowatts, Akosombo 0.36 kilowatts. Others are even worse: the Kompeinga Dam in Burkina Faso generates just 0.28 kilowatts per acre, and Brokopondo in Suriname is worst of all, at 0.08 kilowatts.

Dams' destruction of ecology has been extensive. The commission found that far from "greening the desert" as promised, many dams may have encouraged its advance by desiccating wetlands and delivering salt to fields. A quarter of the world's irrigated land, much of it watered by dams, has been damaged by salt and waterlogging. Meanwhile, accumulations of silt have been reduced by more than half the storage capacity behind a tenth of older dams. And by stopping the flow of silt downstream, dams have universally reduced the fertility of floodplains and "invariably" caused erosion of riverbanks, coastal deltas, and even distant coastlines. Coastal lagoons are being washed away all along the West African coast owing to dams.

A study published early in 2005 found that the world's wild rivers are rapidly becoming extinct. Of the three hundred largest river systems, almost two hundred, including all of the twenty largest, now have dams on them—and engineers are rapidly moving onto the last untamed flows to tap their waters for hydroelectricity or irrigation.

Most of the remaining untamed rivers are in the empty Arctic tundra and northern boreal forests. The largest surviving wild river system is the Yukon in remote northern Canada, the world's twenty-second largest river by volume. Europe's last three undammed river systems are all in northern Russia. But Christer Nilsson, a landscape ecologist at the University of Umea in Sweden, who led the research, said that many dry parts of the world have become "totally devoid of unaffected river systems." And surviving rainforest rivers are under increasing attack. Rivers about to be dammed for the first time include the Salween, which runs for 1500 miles from China through the jungles of Burma and Thailand; the Rajang, in Malaysian Borneo; the Jequitinhonha, in Brazil; the Ca, in Vietnam; and the Agusan, on the Philippine island of Mindanao.

By interrupting natural river flows, dams have wrecked fisheries from the Columbia to the Ganges. Fish stocks established in the reservoirs behind dams rarely come close to compensating. The Grand Coulee and its fellow dams on the Columbia River have destroyed one of the world's largest and most lucrative salmon fisheries. The fish, according to one study submitted to the commission, would have been worth more than the electricity generated by the dams.

The commission report trashed some of the most treasured icons of the industry. Typical was the billion-dollar Kariba Dam, built by the British on the Zambezi River in 1959. It created the largest manmade lake in the world on a rich floodplain where 57,000 Batongan people had lived. It was intended to generate hydroelectricity to boost the future independence and economic development of the peoples of Zambia. But far from benefiting from the project, the Batongans found themselves expelled and left destitute in refugee camps while the electricity and water from the dam went to multinational corporations running copper mines. Was this development? The commission doubted it.

It also catalogued how the Manantali Dam on the Senegal River in West Africa eliminated floods that had provided free irrigation for half a million farmers, and how the Akosombo Dam in Ghana flooded an area of fertile farmland the size of Lebanon while providing a paltry amount of power sold at knockdown price to an American aluminum-smelting company. The loss of silt downstream of the dam has caused massive erosion at the mouth of the Volta River and along the West African coastline. Some 10,000 people who lived along the coast of neighboring Togo have seen their homes disappear beneath the waves.

The commission chairman, Kader Asmal, did not in the end condemn dams as a technology. "Shortcomings are not automatic and can be avoided," he said. Future dams could be justified if they showed that they overcame the drawbacks and had the consent of the people affected. But even so, his report revealed that there are just too many horror stories for each to be dismissed as arising from problematic local circumstances, whether hydrological, social, or political. It was an extraordinary condemnation of what, for much of the past fifty years, has been one of the world's largest construction industries and a cornerstone of world strategies for economic development, soaking up an estimated $2 trillion.

I admit to having a sneaking love of large dams. The sheer grandeur of them takes the breath away. I can empathize with Franklin Roosevelt, who said on opening the Hoover Dam in 1935, "I came, I saw, and I was conquered." With Woody Guthrie, who felt the same way about the Grand Coulee. And with

Jawaharlal Nehru, who called his dams "the new temples of India, where I worship." But clearly dams do both good and harm. A balance, as the commission insisted, can be drawn. But it is not an easy balance.

Usually the benefits are short-term while the costs are long-term. Usually urban elites gain most, especially from hydroelectricity, while the poor in the countryside lose most, as their fields are flooded or the rivers and wetlands on which they depend are wrecked. Water, as they say in the American West, flows uphill to money. That is why farmers and fishermen everywhere in the world are being forced to give up their water resources to cities and industry.

It is perhaps no surprise that, notwithstanding the idealism of the early days, autocratic, corrupt, and militaristic governments have come to like dams best. Marshall Goldman, the analyst of Soviet Russia, identified "an almost Freudian fixation . . . Nothing seems to satisfy the Soviets as much as building a dam." Communist China, with 22,000 large dams, has almost 50 percent of the world total. Spain, under the Fascist leadership of General Franco, built more dams than any nation of comparable size on earth. If nothing else, dams have proved an exceptionally effective technology for turning the unruly flow of rivers into private or state property.

Can we ever draw the balance? At the World Water Forum in Kyoto in 2003, I heard hundreds of scientists delivered a string of serious indictments of many water-engineering technologies, especially dams. They also made numerous proposals for less ecologically damaging and socially divisive solutions to water problems, which we will explore later in this book. And yet water ministers attending the same meeting, but apparently hearing a different message, almost unanimously supported calls for doubling the planet's stock of large dams during the twenty-first century. The simple image of the giant dam delivering water and power and control over rivers seduces them still. It begins to look as if Beard's fond hope that the era of dam building is over may be quite wide of the mark.

The most persuasive mantra is that poor nations must have dams if they are to grow rich. The latest rhetorical version of this claim has been doing the rounds of ministers and international aid officials ever since being road-tested by the dam-building industry in Kyoto. A typical version came from Kristalina Georgieva, the environment director at the World Bank. "Look," she told me during a visit to London, "countries need infrastructure, and that includes dams. Ethiopians have 13,000 gallons of water storage each; Australians, with a similar climate, have almost 800,000." And Americans, she might have added, have double that.

But this is sloppy thinking. Some developing countries might be best advised to build dams, but it is far from self-evident. The antidam campaigner Paddy McCully, head of the International Rivers Network in California, says that such an analysis "does not explain why Ethiopia is poor, or prove that building lots of dams will end its poverty. Zambia and Zimbabwe have around twice the water-storage capacity per person that Australia has, but they remain desperately poor. Paraguay generates nearly ten times more hydroelectric power per capita than Australia but has a per capita GDP just a tenth as big."

A list of the twenty-five countries most hooked up to hydroelectricity includes such economic underperformers as Burundi, Rwanda, both Congos, Malawi, Afghanistan, and Laos. The only wealthy nation among the twenty-five is Norway. I was struck, too, by the situation of the tiny landlocked southern African state of Lesotho. It is one of the most water-rich nations in Africa, with the tallest dam on the continent. That dam has enough reservoir capacity to give each of the country's 2 million citizens about 400,000 gallons of water a year. But in early 2004, Lesotho faced famine as parched crops withered in the fields. The government appealed for food aid. Why? Because almost all the water stored in the mountain kingdom's two giant reservoirs was earmarked for sale to its neighbor, South Africa. The problem for Lesotho was not the absence of water or even of a dam; it was an absence of money.

16

Sun, Silt, and Stagnant Ponds

You could fill every faucet in England for a year with the amount of water that evaporates annually from the surface of Egypt's Lake Nasser. This is one of the more staggering statistics I uncovered while researching this book. According to Egyptian hydrologists, between 8 and 13 million acre-feet of water disappear from the surface of the great reservoir behind the Aswan High Dam— that is, a quarter of the average flow of the river into the reservoir, approaching 40 percent in a dry year.

This huge waste of water in a country dependent for its survival on the Nile should come as no great surprise in Cairo. In the early twentieth century, British colonial engineers forcefully advised against building a dam in the Nubian desert for precisely this reason. It would be far better, they said, to capture water farther upstream, in Ethiopia, where most of the river's water rises. There, a cooler, cloudier climate would reduce evaporation rates and the steep valleys would reduce the surface area of the reservoir. But in the 1950s, when Egypt gained independence and could do what it liked, its leader, Gamal Abdel Nasser, decided on a dam on Egyptian territory, whatever the water losses. He created the second largest manmade lake in the world, with water spreading for 300 miles along the flat Nubian desert deep into Sudan.

Lake Nasser is not alone in such losses, of course. Right across the tropics and beyond, evaporation is a major drain on reservoirs. In the American West as in Egypt, more than 6 feet of water evaporate annually from reservoirs like Elephant Butte, on the Rio Grande, and Lakes Mead and Powell, on the Colorado. A tenth of the flow of the Colorado River evaporates from Lake Powell alone. A typical reservoir in India loses 5 feet. In the parched Australian outback, the losses can exceed 10 feet a year.

What proportion of the reservoir's water supplies this represents depends on the ratio between surface area and capacity and the amount of time the water spends in the reservoir. Like the Aswan High, the Kariba Dam on the Zambezi loses around a quarter of its annual inflow, which is more water than is consumed annually in Zimbabwe. In Namibia analysts estimate that the amount of water evaporating from the Epupa reservoir annually could supply the capital, Windhoek, for forty-two years.

Among the worst offenders must be the Akosombo Dam in Ghana, which holds back a larger surface area of water than any other—approaching 4000 square miles. An evaporation rate of 5 feet would lose 12 million acre-feet a year, which is roughly half the reservoir's input from the Volta River in an average year. Rainfall into the lake may compensate for some of that, but the rain would fall anyway, whereas the evaporation would not happen without human intervention.

Could this evaporation be prevented? Technologists have come up with an ultra-thin layer of organic molecules that can reduce evaporation on small reservoirs by up to a third. But on larger reservoirs, wind breaks the thin surface. And the ecological effects of cutting off the exchange of gases such as oxygen and carbon dioxide between air and water remain largely unknown. So for now we are stuck with the problem.

Igor Shiklomanov, a Russian hydrologist, estimates that 1 quart in every 20 drawn from rivers for human use disappears in reservoir evaporation. That works out at 285 million acre-feet a year, of which 40 percent is lost in Asia, a quarter in Africa, and a sixth in North America. In Australia, 20 million people lose something like 3.2 million acre-feet of water a year—or 53,000 gallons a head. Peter Gleick, the author of a biennial report on the world's water, estimates that an average U.S. hydroelectric dam loses a third of an acre-foot of water per year for every person supplied with electricity. Of course, reservoir evaporation may cause rainfall somewhere else. But even so, it is a lot of water to mislay.

Fetid, choked with weeds, and swarming with mosquitoes, the Balbina reservoir in the Amazon rainforest 100 miles north of Manaus is a billion-dollar boondoggle. The dam rises 150 feet above the forest floor on the Uatuma River, a tributary of the Amazon. The reservoir floods an area forty times the size of Manhattan, but much of it is less than 15 feet deep. Philip Fearnside, from Brazil's National Institute for Research in Manaus, has counted 1500 islands and "so many bays and inlets it looks rather like a cross-section of a human lung."

Even the introduction of a herd of grazing manatees has failed to stanch the spread of weeds across the surface. Stagnant water slips through the reservoir's flooded forest for years before reaching the dam's hydroelectric turbines, which have a paltry generating capacity of 112 megawatts. The reservoir needs to flood the equivalent of a soccer field to deliver enough power to run a small air-conditioning unit back in Manaus.

Even in the Amazon rainforest, that sounds like a waste of land. But the true insanity of this hydroelectric dam has only recently emerged. The rotting vegetation in the flooded forest is producing huge amounts of methane, one of the greenhouse gases thought to be responsible for global warming. The reservoir was created in the 1980s to provide pollution-free electricity for the capital of the Amazon, but by Fearnside's calculations, it produces methane with eight times the greenhouse effect of a coal-fired power station with a similar generating capacity. This is not green electricity.

Balbina is not alone. Brazil is largely powered with hydroelectricity, and Marco Aurelio, of Cidade University in Rio de Janeiro, says that up to half of Brazil's hydroelectric reservoirs warm the planet more than an equivalent fossil-fuel power plant. The World Commission on Dams warned that greenhouse gases bubble up from every one of the reservoirs in the world where measurements have been made. While probably only a handful are emitting more than fossil-fuel power stations, almost all make a significant contribution to atmospheric concentrations. "There is no justification for claiming that hydroelectricity does not contribute significantly to global warming," the commission said.

Vincent St. Louis, of the University of Alberta in Canada, has tried to calculate the global effect of all this methane. He says that reservoirs produce a fifth of all the manmade methane in the atmosphere and make up 7 percent of the manmade greenhouse effect. That is a bigger impact than, for instance, aircraft emissions. His calculation is controversial but persuasive. Until recently, scientists investigating the phenomenon believed that the gases came mostly from vegetation trapped underwater when the reservoir filled. They reasoned that the rotting vegetation would soon be gone and emissions would cease. Not so, it turns out. As reservoirs age, most continue to produce substantial quantities of methane.

Why? For one thing, rotting can be very slow. It takes up to five hundred years for a tree to rot in a stagnant Amazon reservoir. For another, a lot of the rotting vegetation does not come from the reservoir but floats down the rivers that drain into the reservoir. As long as the reservoir continues to flood, the vegetation will continue to arrive and the reservoir will continue to give off greenhouse gases. Of course, most of this vegetation would have rotted anyway. But without reservoirs, says St. Louis, the decomposition would most likely occur in a well-oxygenated river, producing carbon dioxide—whereas tropical reservoirs usually contain little oxygen, and as a result they generate methane instead. Methane is twenty times more potent as a greenhouse gas than carbon dioxide. Reservoirs thus change the way significant amounts of the earth's vegetation rots, and with it dramatically raise the greenhouse effect of the rotting.

This could be political dynamite when the Kyoto Protocol on climate change is discussed. The inclusion of reservoir gases would transform the estimated emissions of greenhouse gases for some nations. If St. Louis's average figure for emissions from tropical reservoirs holds for the Akosombo Dam, the reservoir must be emitting five times as much greenhouse gas annually as all of Ghana's fossil-fuel burning.

The tiny South American nation of French Guiana is in a similar position. Once thought of as one of the world's most greenhouse-friendly nations, French Guiana has a small population, and its industrial emissions are minuscule. But a new dam built in the jungle to power the launch site for Europe's Ariane space rocket is a greenhouse boondoggle on the scale of Balbina. It produces three times as much greenhouse gas as an equivalent coal-burning power station. As a result, French Guiana's real per capita emissions of greenhouse gases are three times those of France and greater even than those of the United States.

———

Reservoirs are not permanent structures—or not permanently useful, anyway. However well they are managed, they accumulate silt brought down from the headwaters of the rivers they trap. The extreme case is the Yellow River, the world's siltiest river, which succeeded in filling the Sanmenxia reservoir in just two years. China's dam engineers are more sophisticated about how they manage silt flows these days, but the reservoirs on the river currently still have a half-life of less than two decades.

No other river beats this, but most rivers flowing out of the Himalayan heartland of Asia carry substantial silt flows. Many dams on these rivers are likely to be as good as useless in forty or fifty years. The reservoir behind the Tarbela Dam, the largest and most upstream on the Indus River, is now more than a quarter full of silt. An island of silt has already broken the surface and is moving ever closer to the dam itself. By 2025 the reservoir will be three quarters filled and effectively useless. Pakistan is looking for a site for a replacement.

Overall, the world's reservoirs are thought to be losing their storage capacity at a rate of at least 1 percent a year. In China, which has more reservoirs than any other country, the figure is 2 percent. That is a loss of millions of acre-feet a year. "This loss should be of the highest concern for governments across the globe," says Rodney White, of HR Wallingford, a British consulting firm. "The world requires between three hundred and four hundred new dams every year just to maintain current total storage."

In theory, clogged reservoirs could in future be replaced with new ones. The trouble is that most of the world's best potential sites for dams are already taken. Any replacement for the Tarbela will be only a fraction as effective. Future dams will get second and third and fourth best, with less water storage and less hydroelectric potential. And their ecological impacts are likely to be much greater. However good the current crop of dams have been for the world, their successors will flood much more land for far less benefit—and be very substantially more expensive.

17

Dams That Cause Floods

Sofia Pedro gave birth to her daughter Rositha in a tree. She couldn't go home or to a hospital because everywhere for miles around her tree was flooded. It was her only refuge. In February 2000, Mozambique had suffered the worst flood in living memory. As water engulfed the country's lowlands, torrential rains got the blame for the inundation. But matters were made worse by the operators of dams upstream in South Africa, Botswana, and Zimbabwe, who had kept their reservoirs almost full at the start of the rainy season.

When the floodwaters came down the rivers, dams that promised to provide flood protection became a lethal liability. Dozens of dams were washed away. The resulting floodwaters came as a surprise to people downstream, who thought the dams protected them.

Most dams are built with the promise that they will capture floodwaters from the rivers they barricade. But one of the secrets of dams is how often they make floods. This happens because of the contradictory hydrological requirements of the different uses to which dams are put. In order to catch floods, reservoirs need to be kept empty, whereas for most day-to-day operations, like feeding irrigation systems, providing urban water, and generating hydroelectricity, they need to be kept as full as possible. Sensitive dam managers try to reconcile these aims. But, says Bryan Davies, a river ecologist from the University of Cape Town who watched the buildup to the Mozambique floods in 2000, "The tendency everywhere is to store like hell when you can."

In times of very high rainfall, a dam's capacity to capture water becomes a menace. As the reservoir fills to the brim, its operators are faced with the choice of risking the catastrophic failure of their dam if it overfills or making emergency releases of water down spillways. Often they make the fateful decision too late. The result is releases that are far greater and more sudden than would happen during natural river flooding.

Nobody keeps proper records, but it is increasingly clear that inept dam management is contributing to a growing toll of flood catastrophes around the world. And many fear "the big one"—a major disaster in which thousands die after a dam bursts or panicking operators release a huge volume of water to save their pride and joy. The risks are greatest when natural sponges on river catchments like forests and wetlands are being destroyed, when dams are designed with poor knowledge of the history of floods on the rivers they block, and when climate change is making all hydrological statistics useless anyway. All three applied in Mozambique.

———

Because nobody knows what the "worst flood" might be, dams are often built without adequate spillways for emergency releases. A World Bank survey in India in 1995 found two dams that could cope with only one seventh of the possible peak flows. For these, said the report's author, William Price, "the consequences of dam failure during a major flood would have to be described with some adjective beyond disastrous." Perhaps he had in mind India's worst dam failure, in 1979, when the Machu II Dam in Gujarat drowned two thousand people.

One of the Indian dams the bank labeled a disaster waiting to happen was the Hirakud Dam, on the Mahanadi River in Orissa. When Sir Hawthorne Lewis, the British colonial governor, laid the foundation stone back in 1946, he promised that the dam would provide a "permanent solution" to floods on the Mahanadi. That was wrong. The dam's managers regularly kept the reservoir too full, and ever since, far from disappearing, serious floods in the river's delta have more than doubled. In July 2001, when engineers made emergency releases from the dam, eighty people drowned and two hundred villages were inundated. Two years later, more than sixty died.

The failings of Indian dams are not unusual. In 1998, the *China Daily* reported engineers' fears that thousands of that country's dams were at risk of catastrophic failure. Quoting unnamed experts, it said that "most of the old dams built in the '50s and '60s are seriously deformed [and] are all threatened by the hidden danger of dam collapse." It suggested that politicians were more eager to build "new star dams dazzling people's eyes" than to repair old ones, which "have been forgotten."

In half a century, 322 Chinese dams had failed. Failures included the world's worst dam disaster, which happened one hot August night in 1975. The operators of the 400-foot-high Banqiao Dam in Henan Province in central China believed that they had nothing to fear when a typhoon hit the hills behind their dam and the Ru River swelled. Their manuals said that the dam could survive a once-in-a-thousand-years flood on the river. But unknown to them, another dam upstream was having trouble. Shortly after midnight, the upstream dam burst, emptying 97,000 acre-feet of water down the river. When the water reached the Banqiao Dam, the dam gave way under the onslaught, unleashing over 400,000 acre-feet of water, mud, and masonry downstream.

One woman, the Chinese later reported, leapt clear with the cry, "The river dragon has come." The flood formed a wall of water 7 miles wide and 20 feet high, traveling downstream at more than 30 miles an hour and crashing into the town of Huaibin with the force of a tsunami. Estimates of the death toll that night vary from 80,000 to 200,000 people. But such was Chinese secrecy back in the 1970s that it was years before the outside world learned of this man-made disaster.

———

Hurricane Mitch hit Central America in October 1998. It was the most destructive storm in the Western Hemisphere in two hundred years and dumped record amounts of rain onto steep hillsides saturated by a month of heavy rain. Landslides and flash floods turned small mountain streams into torrents that killed at least 10,000 people. But many died because of the extra force and suddenness of torrents unleashed as dams collapsed.

Records of the night Mitch struck Tegucigalpa, the capital of Honduras, are less than complete. But when I visited the country six weeks later, I met many people who described how a wall of water had struck poor, low-lying shanty areas in the south of the city shortly before midnight. Several hundred died there. Scientists from the U.S. Geological Survey discovered that minutes before the wall of water struck, the Los Laureles reservoir's dam had burst, unleashing floods along the Guacerique River as far as the city's southern suburbs. Then, on another stream south of the city, the Concepción Dam overflowed, causing what the operators' log called a "high-flow release" through the heart of the capital that lasted for seven hours.

Similar explanations may lie behind mysterious walls of water that hit other Honduran towns that night. I visited Pespire, a town of 30,000 people in the south of the country. It suffered forty-one dead and lost more than three hundred houses after the level of the river that runs beneath the town rose by some 50 feet within a matter of minutes. Nobody could remember anything remotely like it from the normally placid river. But I noticed that the town was just downstream from a dam that would normally have impounded floodwaters. The dam was physically intact. Had its operators made emergency releases? Did the dam trigger the disaster in Pespire that night?

—

In the chaos of heavy rains and already swollen rivers, it is often far from clear where nature's influence ends and humans' begins. Even so, it seems odd that nobody has tried to assess the scale of disasters caused by flood releases from dams. Alerted by what I had heard in Honduras, I carried out a survey of media reports of dam-related flood incidents around the world over the three years after Hurricane Mitch. What I found was, I think, at least a cause for concern.

A year after Mitch, the annual rains were unusually heavy in West Africa. Dams were in trouble. Engineers in the landlocked state of Burkina Faso opened spillways to relieve water pressure on the Bagre Dam on the Nakambe River. Downstream, across the border in northern Ghana, forty-eight people died during the resulting sudden river surge.

Later in the month, large areas of central Nigeria were submerged after operators opened the spillways of Nigeria's giant Kainji Dam, unleashing part of the contents of its 80-mile reservoir down West Africa's biggest river, the Niger. It later emerged that the operators had been jittery about the thirty-year-old dam since it had become too full and almost collapsed during floods the previous year. If that had happened, it would have washed away the town of Kainji. So when the floods returned in 1999, they took no chances. They opened the spillways. Sixty died, and 80,000 refugees needed food and shelter, as their homes and fields were under water.

The following month, on the other side of the Atlantic, media reports held that releases of water from La Esperanza Dam in the Mexican state of Hidalgo were implicated in floods that left a hundred dead. No follow-up study seems to have established the truth of these claims.

Then came the Mozambique floods, when rivers such as the Limpopo and the Zambezi broke their banks, flooding farmland over a wide area of southern Africa. Nobody would claim that dams were responsible for all, or even most, of the flooding. The rainfall was truly exceptional. For three months the Limpopo Valley, which drains out of the African interior, received five times its usual rainfall. At the height of the floods, enough water flowed down the Limpopo in just eight days to fill all twenty-four dams in the river's catchment five times over. But, said local hydrologists, if the reservoirs had been left empty before the rains, the capture of two days' worth of floodwaters could have made all the difference in some places.

Events were equally confused farther north, on the Zambezi, the largest river in southern Africa. The lake behind the Kariba Dam is more than 150 miles long. Through the wet weeks of early 2000 it filled rapidly, until engineers relieved the pressure on the dam by opening three of its five spillways. Initially there was much local excitement. People had never seen anything like it and rushed from miles around to watch. But some 12,000 people living along the riverbank below the dam saw the water surge through their fields, grain stores, schools, and villages. The Mozambique town of Zumbo, 150 miles away over the border, was flooded.

The Zambezi River Authority argued that "if the Kariba Dam had not been there, the areas downstream would have been subjected to worse floods." That may be true. But Kariba's builders had promised that the dam would protect communities from floods. Taking them at their word, tens of thousands of people had moved onto the floodplain below the dam. The authorities obviously took a similar view of the protection offered by the dam. As the BBC reported at the time, "There seems to be no contingency plan in place to help people or businesses... A year's supply of food for thousands of people who farm along the banks of the Zambezi was wiped out in just nine hours."

In August 2000, monsoon floods in the Indian state of Andhra Pradesh cracked the Roxsagar Dam. Thousands of laborers rushed in to mend the breach, but they retreated as more water gushed through and surged toward the state capital, Hyderabad. Nearly 40,000 people fled as water filled their homes.

A month later and a thousand miles north, in West Bengal, engineers made emergency releases from several dams during a flood that left more than a thousand dead. The worst flooding happened around the town of Murshidabad, through which dam releases on the Mayurakshi River flowed, and on the lower reaches of the Damodar River, where at the height of the floods 40 percent of the water came from emergency dam releases. The floods soon spread into Bangladesh. Wide areas of low-lying land along the border disappeared under 10 feet of water. In Bangladesh, water officials said that only those rivers coming from India were in flood. "We consider this flood unnatural and unprecedented," said the director of the country's Water Development Board.

Every year there are more such incidents. In August 2001 over a hundred people died in northern Nigeria after water was released without warning from two dams and swept away two hundred villages. The following year, twenty people died in Syria when a dam on the Orontes River burst, burying more than 39 square miles of farmland in river mud. And in 2003 it was Nigeria's turn once more. At least thirty-nine died after the state power company again sought to save a dam by opening spillways. The incident, on the Shiroro Dam, was a virtual repetition of an incident at the same dam in 1999.

In February 2005, more than a hundred people drowned after a thousand-foot-wide dam burst in heavy rains on the Shadikor River in western Pakistan —just two years after a previous dam on the same site had been washed away. The following month dozens more died when a dam burst in southeastern Afghanistan. And in April, on the Narmada River in western India, more than sixty pilgrims camping on the river's banks were washed to their deaths when managers of the Indira Sagar Dam upstream released water without warning.

Considered individually, each case might be put down to bad luck or poor management. Cumulatively, however, they tell a different story. When they are needed most, dams designed and promoted to prevent floods often end up creating floods.

VI

When
the rivers
run dry...

men go to war over water

18

Palestine: Poisoning the Wells of Peace

Ahmad Qot is a poor Palestinian farmer. He spends between three and four hours every day walking his donkey to get water for his nine children and five farm animals. I met him by chance at a shaft to an ancient tunnel that channels water from the hills above Madama, his village on the West Bank, near Nablus. He repeatedly dropped his bucket down the shaft and hauled it up, gradually filling a long line of plastic containers with water. His donkey stood patiently waiting to be laden with the containers. "I come here three times a day," he said. "I have three cows and two sheep; this is the only water for them."

His life is breathtakingly spartan. He has less than 2 acres of land. He grows a little wheat, to make bread, and some vegetables. His sheep provide his family with meat. His cattle produce milk, which he sells in the village. And the lifeline for his animals is a shaft in the road—a shaft with a beautifully made stone surround that has been deeply grooved by centuries of ropes hauling buckets. Centuries of people getting by as Qot is doing.

The villagers say that the shaft and tunnel are Roman, which means no more than that they are very old. The tunnel concentrates seeps from tiny springs in the hills and brings their flow to the village. Similar tunnels are to be found all across the West Bank. Hydrologists call them spring tunnels. They are an ancient water-catching system that even many Palestinians know little about. Many have dried up as water tables have fallen, but the tunnel at Madama runs all year.

Qot and a handful of the poorest in Madama still use it to water their

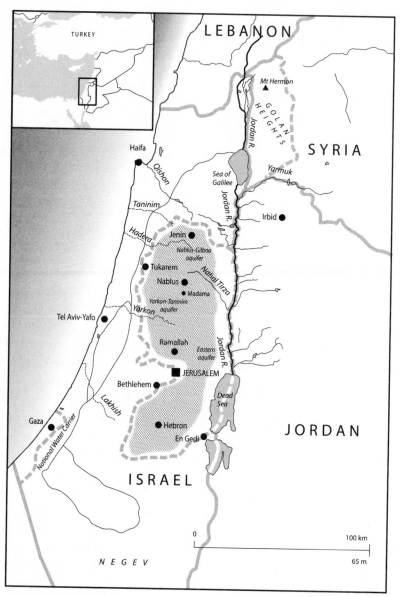

LEBANON

Mt Hermon ▲

G
O
L
A
N

H
E
I
G
H
T
S

SYRIA

Haifa ●

Qishon

Jordan R.

Sea of Galilee

Yarmuk

Taninim

Irbid ●

Hadera

Jenin ●

Nablus-Gilboa aquifer

Tukarem ●

Nahal Tirza

Nablus ●
● *Madama*

Yarkon-Taninim aquifer

Yarkon

Tel Aviv-Yafo ●

Ramallah ●

Eastern aquifer

Jordan R.

■ JERUSALEM

Bethlehem ●

Dead Sea

Lakhish

Gaza ●

National Water Carrier

Hebron ●

En Gedi ●

JORDAN

ISRAEL

0   100 km

65 m

N E G E V

Jordan River and West Bank Aquifer

animals, but lately it has become polluted. "Sometimes even the donkeys won't drink the water," Qot said. Certainly his family cannot drink it. So most days Qot makes another, much longer journey out of the village, across a road that leads to a nearby Jewish settlement, and past a military checkpoint to a spring in the neighboring village of Iraqburin. The journey takes two or three hours —much longer if the soldiers hold him up.

"Some days the soldiers let me pass with my donkey, and some days they don't," he said. "Sometimes when I am coming back, they just pour the water on the ground. They never say why. Once they just drove a tank over my water containers." He cannot understand such behavior. "Water is life," he explained. He has no choice but to make the journey, though. He cannot afford the two dollars for 265 gallons of water charged by the driver of the village's tanker, who also collects water from Iraqburin and delivers to the wealthier inhabitants of his village.

The reality of daily life for Palestinian villagers on the West Bank is told in such stories. It doesn't explain the continuing poisonous dispute between Jews and Palestinians in these hills, but it does go to the heart of the daily misery the dispute brings.

———

The two thousand inhabitants of Madama used to have another source of water. There is a spring two thirds of the way up another hillside—two thirds of the way toward an Israeli settlement, high on a ridge, called Yitshar. A pipe connected to the spring takes water down to a tank outside the Madama mosque. This was the village's main source of clean and free water until the spring became polluted. Everybody in the village assumes that the pollution comes from the Israeli settlement—and I could see no other likely cause. Neither could Geoff Graves, from Oxfam, my guide in the village. "We have twice repaired the head of the spring after Yitshar settlers damaged it. The last time we fished soiled diapers out," he said.

Ayed Kamal, the head of the Madama village council, told me, "Sometime the settlers shoot at us when we go up the hill. Three of us have had bullet wounds. They killed a donkey." Graves said that the settlers, who number between seven hundred and a thousand, have even shot at Oxfam workers repairing the spring. The settlers on the ridge seem to have decided that they are in danger from potential terrorists if they allow the villagers to climb the hill to their spring. But seen from below, their actions are simply vindictive.

Kamal also works as a teacher in the village school, where the outside walls are daubed with pictures of Islamic martyrs and a huge mural of a giant lake surrounded by palm trees. In Madama, he says, getting water is a constant problem. Most families have installed cisterns to collect the winter rains from their roofs and store it underground. "But this water doesn't usually last beyond about May. After that, we all have to buy water delivered by tankers if we can afford it, or bring it by donkey if we can't," he said. But sometimes you cannot even pay for water. The Friday before my visit, a curfew imposed by soldiers on the settlement road prevented the water tanker from coming in.

There is no lack of outside concern over the situation in villages like Madama. Aid flows in. Oxfam has built a water-recycling system at the school. Thanks to its efforts, a series of pipes collects water from the school sinks and pumps it into a tank on the roof, where it flushes a newly repaired toilet block for girls. (Before, girls either went home to use the bathroom or didn't go to school at all.) But despite such projects, the villagers feel hemmed in—by the unknown settlers on the hill above, by the soldiers on the road, and by the many other checkpoints that can make a brief trip to Nablus, just 3 miles away, into a twelve-hour odyssey.

My host advised against trying to visit Yitshar, so I Googled it. The first two references said it all, I imagine. The first, put out by the Israeli Terror Victims Association, listed two people, Shlomo Liebman, aged twenty-four, and Harel Bin-Nun, eighteen, who had been "shot and killed in an ambush by terrorists while on patrol at the Yitshar settlement" on August 5, 1998. The second, from the Israeli Information Center for Human Rights in the Occupied Territories, was a testament by a local Palestinian who had been forced to abandon his olive groves by settlers from Yitshar.

There is clearly mutual fear and hostility here. But while in Yitshar there is running water at all times and sprinklers in the gardens, in Madama times are tougher. Had the villagers ever personally met the settlers? I asked. No, they said, laughing. "They just arrived on the hill. They have never come to introduce themselves."

Most Palestinian villagers on the West Bank tell similar stories of conflict in which water plays a central role. I visited Deir Sharaf, another village outside Nablus, where inhabitants complained that two Israeli settlements overlooking the village have taken their water. Farther east, Beit Fureik and Beit Dajan are two overgrown villages with a combined population of 15,000 but no piped water. People there rely on old wells and more than a hundred rooftop cisterns, installed with aid from the UN Development Program to store the winter rains. In summer, tankers bring water through the security checkpoints that dot the roads to the floodlit hilltop settlements, many of them built on the olive groves of the villagers. Here too Palestinians get shot at if they try to tend their remaining orchards. And here too the Israelis have reason to fear. A Palestinian who blew himself up at a bus stop in Tel Aviv on Christmas Day 2003 was from Beit Fureik. The next day the Israeli army entered the village and shot holes in many of the cisterns.

———

The West Bank has few permanent rivers, but the dolomite hills are honeycombed with caves and crevices, where rainwater collects. There are three aquifers beneath the West Bank. The largest is beneath the western slopes and drains toward Israel proper and the Mediterranean. The second, around Nablus, drains north and supplies much of Galilee. The third drains east to the Jordan Valley. Collectively they are called the mountain aquifers. They are virtually the only source of water for Palestinians, and they are at the heart of the conflict between Israelis and Palestinians over water.

In the 1950s, when the Palestinians lived here under Jordanian rule, the West Bank seemed to have ample water. Far more rain was falling and filling the aquifers than the Palestinians needed. The excess water in the western aquifer gushed from springs along the boundary between the West Bank and Israel. Those springs were the sources of two of Israel's largest rivers, the Yarkon and the Taninim, which flowed west into the Mediterranean. But as Israel's population grew, Israelis began to tap the water of the western aquifer by sinking well fields close to the border. Soon they were taking far more of the water under the West Bank than the Palestinians ever had, without actually setting foot on the West Bank.

"By the early 1960s, we were taking about 240,000 acre-feet from the aquifer and the Arabs about 16,000. Between us we were fully utilizing the aquifer, plus a little more," said Ze'ev Golani, who became an Israeli official on the West Bank. The water table in the western aquifer began to fall, and the two rivers died. The Yarkon's bed became an open sewer for the fast-growing communities of the Tel Aviv urban area. So, said Golani, "When Israel took control of the West Bank after the Six-Day War in 1967, we said there should be no additional pumping by anyone—certainly not for agriculture. And not much has changed since." In that way, almost by accident, Israel came to take the lion's share of the western aquifer and, in the name of conservation, decreed that it should take the lion's share forever.

Israeli hydrological rule on the West Bank since 1967 has been absolute and unyielding. Across most of the West Bank, Palestinians have been largely forbidden to sink new wells, and they rarely get permission to replace old ones. Before the Israelis took control, said Clemens Messerschmid, a German geologist working for the British Department for International Development and the Palestinian Water Authority, the West Bank Palestinians had 774 wells. Thirty-five years later, only 321 were still operating. The rest either dried up or were off-limits in areas requisitioned by the Israeli military. Israel even prevents Palestinians from expanding use of the estimated five hundred springs and spring tunnels. Its only concession is to sell water to villagers with piped distribution networks. But they are in the minority and constantly fear being cut off.

As the existing springs and wells deteriorate and their population grows, Palestinians find they have less water per capita than when the Israelis invaded. The World Health Organization warns that poor water supplies are behind rising rates of waterborne diseases among Palestinians. Today each Palestinian typically has less than a quarter as much water as his Israeli neighbor. "We used to bathe and wash clothes every day; now it is every two or three days. We'd love to have water for a garden, trees, even a pool like the Jews do in the settlements," said Siham Khabirat, a mother of six in al-Dhahriyeh. This is hydrological apartheid.

And often Palestinians pay more for their limited supply than Israelis do. Tanker water typically costs two dollars for 265 gallons. A study in the villages around Nablus found that families spend between 20 and 40 percent of their income to buy water. The highest spending was in Salfit district, which has some of the most productive wells in the country. But most, said Messerschmid, have been confiscated by the Israelis to supply their settlements.

Many outsiders believe that Israel annexed the water of the West Bank only when it took over the territory after the Six-Day War. In fact Israel did not have to occupy the West Bank to take the water; it was doing so years before the war. But the subsequent occupation did stop Palestinians from developing their water resources and insured that Israelis could continue to take the lion's share.

The Oslo Accords, signed by both sides in 1995 as part of the peace process, cemented the existing unequal share-out of the mountain aquifers. They allowed Palestinians 15,000 gallons of water per person per year, whereas Israelis get 65,000 gallons each. But the Palestinians pinned their hopes on what they saw as a promise by the Israelis to join them in developing new sources of water in the region, particularly to exploit the eastern aquifer. This had the attraction of being the only water resource in the region entirely within Palestinian borders. "It has been the focus and the hope of the Palestinian Water Authority," said a local British water consultant, Jennifer Moorehead. After the accords were signed, foreign donors came in to help the Palestinians realize their dreams.

But this hydrological peace dividend is proving to be an illusion. There are growing doubts about how much water is in the eastern aquifer, and even more about how much it will be possible to exploit. "The only good places for drilling on the east side are down in the Jordan Valley. But it's just not economical," said Messerschmid. "The water table is a third to a half a mile down, and there have been so many dry boreholes that at a thousand dollars a yard, the aid agencies don't want to drill anymore."

The U.S. government's aid agency, USAID, which has been drilling deep wells to supply Hebron and Bethlehem, agrees. Alvin Newman, the head of water resources at the agency's Tel Aviv office, told me, "We are getting to the end on the eastern aquifer. If you are pumping water half a mile out of the ground and then another two thirds of a mile uphill to faucets in Nablus, that involves very big costs. In fact, you can't buy water pumps big enough to do it. We buy oil-industry pumps."

Most outside experts put the maximum potential of the aquifer for Palestine at 16,000 to 24,000 acre-feet a year, said Moorehead. That is only about a quarter of what was envisaged under the Oslo Accords. The Israelis disagree, in public at least. Yossi Gutman, the chief hydrologist for the Israeli water utility, Mekorot, said, "Yes, there is a problem in some areas, but I believe we could pump water up to Palestinian villagers. Maybe the Palestinians don't want to invest in this project." But if so, they are not alone. And for now, Palestinian water officials deeply regret the concessions they made in Oslo. They don't see how they can supply their people without getting their hands on a greater share of the western aquifer, yet all the time their share seems to be diminishing.

The most recent problem is Israel's security fence, which has cut off dozens of Palestinian villages from their wells. The fence nominally follows the "green line," the internationally accepted border between Israel and Palestine, sealing off the West Bank from Israel proper. But it does not follow the border precisely. The "fence" is in fact a complex *cordon sanitaire* of trenches, barbed wire and electrified fences, turrets, infrared sensors, and a security road. It is up to 325 feet wide and constructed entirely on requisitioned land on the Palestinian side. And in places it plunges for dozens of miles into Palestinian territory in order to keep Jewish settlements on the West Bank on the Israeli side of the barrier. A number of Palestinian villages have found themselves on the "wrong" side of the fence, and others find that although they are on the Palestinian side, their wells have disappeared behind the fence.

Israel says it planned the fence to keep suicide bombers from gaining free access to Israeli towns. But hydrologically, it could not be more divisive. The most productive wells on the West Bank, for both Israelis and Palestinians, lie in a strip of land on either side of the green line. So, said Messerschmid, "The wall has taken many of their best wells from the Palestinians." Most notably, at Baqa ash Sharqiya, near Tulkarem, wells supplying 400 acre-feet of water a year to fifteen villages are now on the Israeli side of the fence, while the villages are on the Palestinian side.

In mid-2004, Messerschmid calculated that wells and springs that once delivered 4000 acre-feet of water a year to Palestinians sat on the wrong side of the barrier. If so, that represents almost a quarter of the water the Palestinians take from the prized western aquifer. Some have concluded that the barrier is a conspiracy to appropriate water. In truth, said Messerschmid, the hydrological implications are probably just an accident of Israeli's security agenda. But that doesn't make them any less painful.

———

The conflict is having longer-term effects on water supplies in the West Bank, which could be catastrophic for both communities. In the fractious frontier lands of the West Bank, few care these days what happens to garbage and sewage. Both the settlers and the Palestinians pour their waste into old quarries or down the banks of wadis. Geoff Graves has seen sewage tankers discharge their toxic load into cracks in the rocks that lead straight to the aquifer. "Month by month, since the intifada and the restrictions on movement on the West Bank, you can see the wadis filling up with sewage," said Messerschmid. "In terrain like this, that means the wastewater will go right into the aquifer." Drivers say that constant checkpoints leave them little choice.

Outside Deir Sharaf, I saw a stream of raw sewage discharged from Nablus into the Wadi Zeimar, which recharges the western aquifer. In Ramallah, the Palestinian administrative capital of the West Bank, the city sewage plant is overloaded and provides effectively no treatment. But aid projects to build sewage treatment plants for these towns have fallen foul of the security crisis.

Though the settlers make up only a tenth of the total population, Friends of the Earth in Israel estimates that they may be responsible for a quarter of the 50,000 acre-feet of untreated sewage that flows into the West Bank aquifers from more than 2 million people of both communities. "The settlements all discharge untreated wastewater into the wadis," said Robin Twite, a former head of the British Council in Israel who is now head of the Israel-Palestine Center for Research and Information.

Israeli boreholes are deep and remain unpolluted. But in many places sewage is already polluting Palestinian springs and wells, causing yet more to be abandoned. At Nazlat'Isa, north of Tulkarem, artesian wells that bring water to the surface under pressure are laced with sewage. "The people know the water is polluted, but they have no choice but to use it," Monther Hind, a Palestinian wastewater engineer, told me in his office in Ramallah. The sewer pipe that was leaking, he said, was on the other side of the security barrier, so the villagers could not repair it.

———

How, ask the Palestinians, are we expected to survive under such conditions? Why are we allowed to have only 20 percent of the water that falls onto the West Bank? The Israelis, even at the height of the intifada, agreed that these were valid questions. But while the Palestinians insist that the answer lies in giving them greater access to the aquifers, it emerged in 2004 that the Israelis have another plan: let them drink seawater.

Israel, which has been developing desalination for its own use, has drawn up a plan for a giant desalination plant on the Mediterranean coast that would be dedicated to supplying a network of pipes across the West Bank. It wants the U.S. government to pay for the plant and the pipes, which could, it says, supply a wide area, from Jenin in the far north of the territory to Ramallah in the south, including many of the 250 villages that currently rely on local springs and small wells for their water. Israel says its contribution would be to guarantee safe passage of the water across its territory. But it would do this only in return for an agreement that it can continue to take its current share of the waters of the West Bank—especially the western aquifer, the region's largest, cleanest, and most reliable water source.

USAID, which would probably fund the scheme, puts the capital cost at $800 million. But the running costs would also be high. Large amounts of energy would be needed both for desalination and for pumping the water up between 1000 and 5000 feet to the West Bank communities that need it.

Is this a viable solution? Uri Shamir, the director of water research at the Israel Institute of Technology in Haifa, told a U.S. House of Representatives committee in 2004 that the desalination project was "the only viable long-term solution" for supplying drinking water to the West Bank. Later, he told me that the project could be completed in five to seven years. "The plant will be funded by the world for the Palestinians," he said. "Israel will not be willing to carry this burden, and the Palestinians are not able to." Alvin Newman at USAID agreed.

But Palestinian negotiators are deeply uneasy about the plans being drawn up on their behalf. "Desalination cannot be a substitute for our rights of access to the western aquifer and the Jordan River," said Nabil Sharif, the chairman of the Palestinian Water Authority. Many outsiders share the unease. Tony Allan, of the School of Oriental and African Studies in London, a leading authority on Middle East water, says that "pumping desalinated water to the West Bank is not the best technical or economic option."

Even leading Israeli hydrologists are incredulous at the plan, which they say would cost more than $1,200 for every acre-foot. "It would be foolish to desalinate water on the coast and push it up the mountains when there are underground water resources up there that cost only a third as much," said Arie Issar, a leading Israeli hydrologist for half a century, who works at Ben-Gurion University of the Negev.

The Israeli water commissioner, Shimon Tal, sees the issue through the prism of domestic Israeli politics. "It is very difficult to take water from some consumers and give it to others," he told me. "We can't solve the problem with natural water. We must start talking about production of desalinated water for the West Bank. If we can cooperate to produce more water resources, then the negotiations on natural resources will be much easier."

When Israel first announced plans to build desalination plants for its own needs, many liberal Israelis thought those plants could reduce Israel's demands on the West Bank aquifers and allow them to be partially returned to the Palestinians. But the hard truth is that the current generation of Israeli water planners sees desalination as a means of retaining control of those aquifers, not of relinquishing them. And with desalination plans also afoot to supply Gaza, they are looking forward to a future in which an independent Palestine becomes more dependent on desalination than almost any other nation in the world.

I spoke to many people on both sides about the potential for redistributing western aquifer water to Palestine. Former West Bank water commissioner Golani believes that Israel makes too much fuss about the western aquifer. "Arab pumping today is only a small fraction of total Israeli use," he said. "If they doubled consumption, it would take perhaps ten years and still be only marginal. They should have a bigger share." Issar thought it would not be unreasonable to increase the Palestinians' share of the water from the mountain aquifers to 50 percent.

But so far the Israeli government has refused to countenance these ideas, even from the heart of its hydrological establishment. It regards the western aquifer as the hydrological crown jewel of Palestine, a more vital resource for Israelis even than the land of the West Bank. Many Israeli leaders believe that they need to maintain an iron grip on the aquifer in order to keep their own water flowing. And if that means keeping control of the West Bank itself to prevent new drilling by the Palestinians, then so be it.

Unless the Israelis are willing to unlock that jewel, then Ahmad Qot will go on dropping his bucket into the Madama shaft—and the prospects for hydrological peace on the West Bank will continue to look dim.

19

The First Modern Water War

In 1964 Israel hijacked the waters of the Jordan River. There is no other way of putting it. The Jordan Valley, a green desert strip that had been cultivated for longer than perhaps any place on earth, was overnight deprived of most of its water. One day the Jordan poured out of the Golan Heights into the Sea of Galilee, and on down the valley to the Dead Sea, as it had for millennia. The next day a dam constructed by Israeli engineers blocked its outflow from the Sea of Galilee. Instead, a pumping station lifted the water out of the sea and into a 10-foot-wide pipeline that delivered it the length of Israel.

Grabbing the Jordan's flow was no easy task. Engineering heroics were involved. By the time the river reaches the Sea of Galilee, it is more than 600 feet below sea level, in a great fissure in the earth's crust known as the Rift Valley. To divert that water to Israelis' faucets, it had to be lifted 1200 feet out of the Rift Valley. But this was accomplished. The pipe is known today as the National Water Carrier. It carries more than 400,000 acre-feet a year and has become the major source of water for all of Israel.

The carrier irrigates fields from Haifa in the north almost to the Egyptian border and fills faucets from Tel Aviv to Jerusalem and the Israeli settlements on the occupied West Bank. Some is even sold to Palestinians living in the shadows of those settlements. The pumps that run the National Water Carrier take an eighth of the entire output of Israel's power stations. At the time of writing, in early 2005, no fresh water had flowed out of the Sea of Galilee into the lower Jordan Valley since 1991. "It is an article of faith among Israeli hydrologists that to let the water into its own valley is a failure," said Gidon Bromberg, of Friends of the Earth in Israel.

The emptying of the Jordan River may have been an engineering triumph, but it was also a geopolitical earthquake. It was done without the agreement of either Syria, which then held the eastern shore of the Sea of Galilee, or Jordan, through which the river ran to the Dead Sea. It was as if France had one day unilaterally annexed the Rhine as it flowed out of Switzerland and pumped its contents over the hills to irrigate the plains of northern France, leaving a dribble of water to flow down Germany's main artery.

But this was not the end of the matter. Almost three years later, Israel and its Arab neighbors fought the Six-Day War. Most histories of this conflict discuss its importance and motivation in terms of land and security. They often ignore water. Yet the simple fact is this: before the war, less than a tenth of the Jordan River's basin was within Israel's borders; by the end, the basin was almost entirely controlled by Israel. Israel had taken from Jordan all the land between its former eastern border and the river. And it had taken from Syria the Golan Heights, the mountains northeast of the Sea of Galilee where the headwaters of the river rise.

Ariel Sharon, who was a commander in that war and much later became prime minister, is unabashed about Israel's hydrological motives in that conflict, though he said the other side started it. In the early 1960s, he wrote in his autobiography, Syria committed the first offensive act on the Jordan River by starting to dig a canal in the Golan Heights to divert the headwaters away from Israel. "The Six-Day War really started on the day Israel decided to act against the diversion of the Jordan," he wrote. "While the border disputes were of great significance, the matter of water diversion was a stark issue of life and death."

The Six-Day War was, by this account, the first modern water war. And Israel's victory in seizing the Jordan River and its catchment remains an essential backdrop to the continuing conflicts. Israel today uses far more water than falls on its territory, and it has been able to do so because of its occupation of the West Bank, which gives it control of the western aquifer, and the Golan Heights, which gives it control of the Jordan River. That is how hard-line hydropoliticians in Israel see things, at any rate. But almost any final peace settlement with Israel's neighbors will require Israel to hand back some of its control and some of the water.

—

Israel now has three main sources of water. The first, used from the day the state was founded after the Second World War, is an aquifer that stretches along the coast from Haifa in the north to Gaza in the south. Ze'ev Golani, an Israeli water engineer and administrator for many years, drilled many wells into the coastal aquifer as the new state established plantations of water-guzzling crops such as cotton, tomatoes, avocados, and the ubiquitous Jaffa orange. Soon, he explained, the water table fell and—as the Palestinians have discovered in Gaza—the porous rocks began to fill with salty seawater flowing in from the Mediterranean. Pumping was cut back, but even today much of the aquifer, which supplies up to a fifth of Israel's water, is too salty for drinking.

So in the late 1950s Israel began to sink wells farther inland, near the Yarkon and Taninim springs. Those wells, as we have seen, tapped the edge of the western aquifer beneath the West Bank. By the early 1960s, the Israelis were overpumping there, too. And that was when they decided to annex the third, and currently biggest, source of water for Israel—the Jordan River.

The fate of the coastal aquifer is essentially a private matter for Israelis. The fate of the West Bank aquifers is a matter of intense dispute between Israelis and Palestinians. But the fate of the Jordan involves many more players, including Syria, Jordan, the Palestinians, and from time to time Lebanon. So much hangs on a river that in strictly hydrological terms is so small. From its sources in the Golan Heights to its former end point in the Dead Sea, it is barely 200 miles long. It drains an area the size of New Jersey and, as Mark Twain once observed, "is not any wider than Broadway in New York." But throughout human history its waters have been vital in this region, and the current battle over who controls them is in deadly earnest. Here, when the river runs dry, war looms.

In all real senses the Jordan River is now two rivers. One begins amid the snowy slopes of Mount Hermon on the Golan Heights, which is legally in Syria but since 1967 has been occupied by Israel. Its main source here is the Dan River, which bubbles to the surface in a spring amid glades of laurel, ash, and fig close to the ancient Phoenician city of Laish. Two other streams, the Banias and the Hasbani, which begins life in Lebanon, join before it crashes down through a narrow gorge into the Rift Valley and the Sea of Galilee and then passes into the netherworld of the National Water Carrier. The true end point of this river today is the Negev town of Mitzpe Ramon, a desolate hilltop outpost housing East European immigrants, where the carrier itself ends. Its water is at this point 3600 feet higher than when it left the Sea of Galilee.

The second Jordan River is the former tributary known as the Yarmuk River. It rises in Syria and enters the bed of the Jordan 6 miles downstream of the Sea of Galilee. The bed is empty here, except for a trickle of saline water from salty springs on the shores of the sea, which Israel does not want to pollute its carrier water. But the Yarmuk too is a shadow of its former self. Syria grabs half its flow in dozens of small dams on the headwaters. Jordan diverts much of what is left into the 60-mile-long King Abdullah Canal, which passes down the Jordan Valley on the east side of the river, irrigating fields as it goes. Even Israel takes some water from the Yarmuk, extracting it from the spot where the river joins the Jordan—a military outpost it grabbed in the final hours of the Six-Day War in order to demolish a half-completed Jordanian dam there.

The Jordan Valley of biblical legend once carried more than 800,000 acre-feet of water a year to the Dead Sea. Now it carries less than a tenth as much. Most of it is sewage diluted with occasional winter storm flows out of Syria. The spot where John is supposed to have baptized Jesus is now a sewage seep behind the barbed wire of a military fence. And soon the dilution could be gone. In 2004, Syria and Jordan announced plans for a joint dam, dubbed the Unity Dam, to capture most of what remains of the Yarmuk's water. That, said Bromberg of Friends of the Earth, "will probably take the last of the Jordan's regular flow"—the last of a river holy to Jews, Christians, and Muslims alike, which has watered civilizations for 10,000 years.

And the Dead Sea? It is famously the saltiest large body of water on the planet. And the lowest. With next to no water entering it from the Jordan Valley, it is becoming saltier and lower. Its surface is now 1368 feet below the level of the world's oceans and more than 80 feet below where it was fifty years ago. It is falling by 3 feet every year. A third of the sea's surface area has gone, and it has divided in two. Meanwhile the retreating shoreline is leaving behind mud and quicksand, which makes it unsafe to walk along much of the shore. And thousands of sinkholes, some bigger than a car, have opened up as salt layers have dried out along the shore and cavities have collapsed. Roads are buckling and bridges are giving way. An area around the town of Engedi has been effectively abandoned. The sea is dead indeed.

———

But hydropolitics are very much alive. Israel remains worried that if it withdraws from the West Bank or the Golan Heights, it will be powerless to prevent its neighbors from cutting off or contaminating its water supplies. It is afraid that Palestinians could ruin the western aquifer and that Syria might poison or divert the waters of the Jordan before it leaves the Golan Heights. Because of these fears, Israel fights harder to defend water even than it does to defend land. In 2000, President Clinton came close to persuading Israel to hand back the Golan Heights to Syria as part of a peace package, but the deal failed, diplomats say, over who should have control of the tiny streams at the head of the Jordan River. Two years later, Israel threatened military action to force Lebanese engineers to stop supplying a few villages with water from the springs at the head of the Hasbani during a drought.

Some in Israel will not give an acre-foot, let alone 1000 acre-feet. "Implementation of proposed peace initiatives would [mean] authority and control over almost 70 percent of Israel's water supply would be transferred to the Arabs," said Martin Sherman, an influential political scientist at the University of Tel Aviv and the author of *The Politics of Water in the Middle East.* "Either Israel has sole control of her national water resources or her very survival is threatened," as one pro-Israeli American pressure group put it.

Survival? How so? Well, Israel sees survival in terms of preserving a Western lifestyle. Sherman put it like this: "Frequent showers, swimming pools, well-groomed private gardens, and public parks all involve water consumption, without which adequate standards of modern life cannot be attained." Ceding water for peace is not, in many Israelis' eyes, an option, because it would threaten that lifestyle. End of argument.

This rhetoric makes it sound as if war is inevitable, as if water is in such short supply here that peace over such a vital resource is impossible. But further water wars are not inevitable, and water is not really in such short supply. Israeli hydrologists, if not their political colleagues, know that they have options beyond fighting with their neighbors for water.

In the past five years, Israel has embarked on a program of constructing desalination works to provide drinking water from the sea. The first opened in 2005. Its water costs little more than the water expensively pumped up from the Sea of Galilee. A national plan agreed to in mid-2004 will raise production of desalinated seawater to half the country's current water demand by the end of the decade, according to Water Commissioner Shimon Tal. Israel also has plans to increase substantially the amount of sewage that it recycles to irrigate its farms. Even so, Tal continued to maintain in 2004 that the country's water needs would rise steeply and that "redistribution of existing water resources [to Palestinians] is not a solution."

Many of his hydrologists disagree with him. They say that no such rise in demand needs to occur. It is patently ridiculous, they say, for Israel to use two thirds of its water to irrigate crops that generate less than 2 percent of the country's GDP, and even to export virtual water in the form of oranges and tomatoes. To take an extreme option, if Israel shut down its agriculture altogether, it could provide large amounts of water for Palestinians while improving its own hydrological security—with barely a blip in its economic well-being. If it wants to maintain agriculture, these hydrologists say, it can develop desalination further for drinking water and aim to use that water a second time by recycling treated sewage onto farmland. That would solve the problem.

⎯

I talked about all this with Arie Issar, a man steeped in the history of Israeli water. More than half a century ago, before there was a Jewish state, his father sold ice with Arab friends in Jerusalem. As a young scientist in the new state of Israel, Issar helped to green the Negev Desert by finding new sources of water beneath the sands. He says there are still more to be found. But he has always preached careful use and sharing of the region's water.

We met one hot summer's day in Jerusalem. We looked out over the arid Jordan Valley, which his fellow hydrological engineers had emptied, and north to where Israeli soldiers prevented Palestinian villagers from sinking wells to meet their most basic human needs. An old Zionist idealist, Issar clearly mourned the new belligerence of modern Israel. He confided that winning the Six-Day War had been in some respects a disaster for his people: "After that, we stopped developing the Negev, which was largely empty, and began to occupy the West Bank, which was already populated by the Palestinians. Now we are fighting over land and water in the West Bank when we could be developing the land and water in the Negev. We need to get back to that idea.

"People talk about water wars, but water can also be the basis for peace," he said. "And I think it can be so here. We Israelis use too much drinkable water for irrigation when farming is no longer important for our economy. We do crazy things like turning fresh water into oranges and exporting them. The Arabs need that water. They should have it."

20

Swords of Damocles

Southern Doda in Indian Kashmir is bandit country. Pakistan claims the region, and Muslim youths regularly cross the border into Pakistan for training, returning as guerrillas to fight Indian security forces. The war has been going on for more than a decade, with an estimated 50,000 dead. It usually hits the headlines only when foolhardy Western tourists get caught up in it. But some people believe that the world's first nuclear exchange could one day be triggered between India and Pakistan because of events up here in the foothills of the Himalayas. If so, it may not be over guerrillas or terrorists or even border incursions by regular soldiers. It could be over water, for Kashmir is the gateway through which Pakistan receives most its water along tributaries of the Indus River.

Pakistan's 150 million people would be in trouble without the Indus. It runs the length of their country. Its waters irrigate most of their crops and generate half their electricity. But the Kashmir gateway is an Achilles heel. India and Pakistan have been in armed conflict three times since the two states were formed on British withdrawal from the subcontinent in 1947, and the first conflict arose when India intervened in Kashmir to cut the flow of tributaries of the Indus on which Pakistan relied. Some in Pakistan fear that India may be about to do the same again.

After a decade of negotiations and skirmishes following the first conflict, the World Bank brokered the 1960 Indus Waters Treaty between the two countries. The treaty bound them to share the flow of the river, with each taking water from three tributaries. The flow of the Chenab River was among those going through Kashmir that were given to Pakistan, and it has since been the biggest source of water for Punjab, the breadbasket of Pakistan. But geography dictates that India always has the potential to stop Pakistan's water. And now, in the heart of southern Doda, Indian engineers are building a dam on the Chenab.

Pakistan sees the 525-foot-high, billion-dollar Baglihar barrage, rising from the bed of the Chenab just before it crosses the border, as a clear breach of the treaty. India denies any breach. The dam is intended only to generate hydroelectricity, it says. It will not remove any water from the river, only channel its discharge downstream through turbines. India says Pakistan will get all the water it should. But Pakistan says that India could, in a future crisis between the two nations, use the barrage to hold up the river's flow during the critical winter planting season, thus causing famine in Pakistan.

Since 1999, when India unveiled its plans, Pakistan has been demanding that the dispute go to arbitration. So far India has refused. Instead it appears to be rushing to complete the dam. The two countries are still talking, but the Indians are still building. And nobody can forget that since the last dispute over these waters, both sides have developed nuclear weapons. Could this beautiful spot be the flashpoint for the unthinkable?

The stakes are huge. The Indus and its tributaries make up one of the world's largest river systems. Most of the water rises in the rain-soaked mountains of the Indian Himalayas and the Hindu Kush in eastern Afghanistan before flowing through Kashmir and reaching arid Pakistan. For almost four decades the Indus Waters Treaty has been held up as a model for other countries with water disputes. But in truth it has been enforced largely at the insistence of the World Bank, which refused to give either country loans for new dams unless it signed up.

Even without the Baglihar imbroglio, tensions over the treaty have been growing. In 2002, during the most recent standoff between India and Pakistan over Kashmir, some nationalist politicians and commentators in India were so angered by terrorist incursions that they talked of unilaterally taking back the Indus tributaries currently allotted to Pakistan. Most vociferous were the

Kashmiri politicians, who want the tributaries that flow through their territory for their own people. Jasjit Singh, the former director of the Institute of Defense Studies and Analyses in New Delhi, threatened, "Under normal circumstances, India would have continued to fulfill its moral obligation of sharing its water with Pakistan. But unusual circumstances call for unusual action."

Pakistan finds itself in the classic position of the downstream state—at the mercy of others for its most basic resource. But from the Indian viewpoint, things look quite different. As Indians see it, Pakistan got an extraordinarily good deal with the Indus treaty. Both Afghanistan and India contribute far more water to the Indus system than Pakistan does but take far less. It is not their fault if Pakistan has hitched its wagon so firmly to the river. Moreover, they grumble, Pakistan uses its share of the river with spectacular inefficiency while suffering no penalty. Having got itself into a position of undue dependence on the river, the Indians suggest, Pakistan would be better off not raising the specter of a water war—a war from which it could only lose in a manner devastating for its huge and fast-growing population.

The whole world should care about this obscure hydrological battle in a part of the world where few dare to go. Though tensions eased after Indian elections in 2004, the rumblings of the modern world's second water war never stop.

⸺

The dispute over the Indus is just one of a series of confrontations surfacing around the world as water shortages increase along shared rivers. A quick look at the atlas suggests the extent of the problem. Most big rivers are shared. A host of upstream nations threaten to grab the water before equal numbers of downstream nations can fill their cups. The Nile passes through ten countries. The Danube, Rhine, Niger, and Congo all pass through nine, and the Zambezi through eight. Almost half the world's population lives in international river basins. Two thirds of these basins have no treaties for sharing their water. Each is the scene for a potential water war.

Many downstream countries are dangerously dependent on external water supplies. Egypt is even worse off than Pakistan. Virtually rainless, it gets

97 percent of its water from its numerous upstream neighbors on the Nile. The country's leaders say they will go to war if any of them starts diverting the river's water. That probably means Ethiopia, where three quarters of the flow begins. The current international treaty on the Nile, a relic of British rule on the river, gives most of the flow to Egypt, a modicum to Sudan, and none at all to anyone else, including Ethiopia. Clearly that situation cannot last forever, but will Egypt cede any of its current entitlement in order to reach peace on the Nile? It has not as yet ceded anything.

Africa—a continent of haphazard boundaries largely created in the days of imperial rule and maintained because anything else would bring chaos—is full of countries dependent on their neighbors for most of their water. It has eighty transboundary rivers. The isolationist Islamic republic of Mauritania, in West Africa, gets 95 percent of its water from beyond its borders, mostly from the Senegal River coming out of Guinea and Mali. Tiny Gambia receives 86 percent of its water from Senegal. And another desert state, Botswana, relies on others for 94 percent of its water. The Limpopo comes from South Africa, and the Okavango rises in Angola and passes through Namibia before draining into a desert delta, a wildlife mecca where tourists contribute a tenth of Botswana's GDP.

Outside Africa, the situation is little better. Bangladesh, like Pakistan, receives more than 90 percent of its water from India, which has built a barrage on the Ganges right on the border. Meanwhile, Syria and Iraq rely for most of their water on the Tigris and the Euphrates, flowing out of Turkey, which is building dams on both rivers. During the first Gulf War, Turkey considered shutting the Ataturk Dam on the Euphrates to put a hydrological squeeze on Iraq, though it never followed through.

All told, more than twenty nations get more than half of their water from their neighbors. Many should by rights get more than they do. Mexico receives virtually none of the flow of the Colorado River out of the United States. Most of the water in the Jordan River disappears within Israel and never reaches the country named after it. The Illi has shrunk by two thirds by the time it leaves China for Kazakhstan. The Karkeh, flowing west out of Iran, rarely makes it to Iraq anymore. Iran, meanwhile, has rarely seen the Helmand flow west over the border from Afghanistan since the 1990s. Given the extent of the damage caused by such disruption to natural river flows, it is perhaps a wonder that there have not been more out-and-out wars.

———

Many water disputes are simmering in places that few people have heard of and, for the moment at any rate, even fewer care about. They are concentrated particularly among the fractured remnants of the former Soviet Union. Moscow had a notable enthusiasm for large engineering projects, especially dams and big water diversions. Many were built in contentious mountain regions held together only by Soviet military, economic, and political might. In those days, Moscow had absolute control over who got what water and when. But with the glue of Soviet rule gone, the keys to the sluice gates are in the hands of new upstream states, which have equally new downstream nations at their mercy. From Armenia to Tajikistan and the independent-minded Russian republic of Chuvash, which some see as the next Chechnya, nasty local disputes have involved threats of hydrological terrorism. Witness the Ruritanian chaos in the Caucasus Mountains, where a giant hydroelectric complex is at the heart of a long-running conflict between ex-Soviet Georgia and the breakaway republic of Abkhazia.

The Inguri complex, which was completed by Russian engineers in 1980, is 892 feet high, making it the world's third tallest dam. It controls a river of the same name. It provides most of Georgia's electricity and is the only source of power for Abkhazia. But while the dam is on the Georgian side, the power plants are in Abkhazia. Through several years of armed conflict, engineers have managed to maintain electricity generation for the benefit of both sides— largely because Russia, which also takes a portion of the power, has insisted that its own troops guard the complex. In effect, the complex has been a small slice of the old Soviet Union sustained on the border between the new warring republics.

But the situation is perennially unstable. Tensions rose in mid-2004 when masked men abducted two engineers near the plant. The Georgian authorities accused Russian special forces of unspecified subversive activities against its power industry. The question of whether Abkhazia becomes an independent republic recognized by the UN or stays within Georgia probably hangs ultimately on who gets control of the Inguri complex—assuming that it survives, of course. The complex has seen only sporadic maintenance since the breakup of the Soviet Union, and it happens to straddle a potentially active fault line that could trigger an earthquake at any moment.

———

Increasingly, too, rivalries over water between different parts of relatively stable countries are overflowing into violence. The states of Karnataka and Tamil Nadu, in southern India, have been feuding over the Cauvery River for decades. It began in 1927, when upstream Karnataka built the Krishnaraja Sagar Dam, creating an 80-mile-long reservoir on the river. Tamil Nadu responded with the Mettur Dam in the 1930s. Both states built extensive irrigation systems and encouraged their farmers to grow water-intensive cash crops such as rice and sugarcane, in efforts to cement their claims to the bulk of the river's water.

Initially this was just economic one-upmanship. But the stakes have risen. Today both states have populations as large as those of major European countries. On top of escalating rural demand, cities such as Bangalore, the heart of India's Silicon Valley in Karnataka, are demanding ever more water for themselves. Estimated demand from the two states has now reached 20 million acre-feet a year. Average flow on the Cauvery is only 16 million acre-feet. Inevitably, the upstream state, Karnataka, is in the driver's seat. Many years it refuses to deliver downstream the water that Tamil Nadu expects and demands. Below the Mettur Dam, the Cauvery is virtually empty. Rice fields on the river's delta have no water.

During droughts, the conflict spills over into riots. In 1992, twenty-five water rioters died in Bangalore. In October 2002, after the worst drought in southern India for forty years, India's Supreme Court ruled that Karnataka must release water downstream. But farmers blocked the state's main highway to protest against the releases. The government responded by shutting schools and colleges and suspending all train services. The state's business virtually ground to a halt as thousands of riot police broke up demonstrations with tear gas and surrounded the main dams to prevent angry farmers from attacking them.

There are important reasons that warmongers with serious designs on power may hold back when it comes to dams. During two decades of guerrilla warfare in northern Mozambique in the 1970s and 1980s, fighters on both sides spared the giant Cahora Bassa Dam on the Zambezi. The reason? They expected to win—at which point they would need the dam. But in the twenty-first century, many terrorists have little desire to take power and few prospects of doing so. Therefore, they have less reason to hold back. And they will know that there are few better ways of inflicting terror than threatening to attack a dam, and few better ways of causing chaos than carrying out the threat.

In September 2004, the Chinese military suddenly sent helicopters, patrol boats, armored vehicles, and bomb-disposal robots to the site of its giant new dam at Three Gorges on the Yangtze River. Nobody is sure what was in the minds of the authorities, but the action followed press stories about the risk of terrorists hijacking a large ship on the Yangtze, packing it full of explosives, and ramming it into the dam. And it came only a few weeks after the Pentagon had reported that the Taiwanese military was considering just such an attack on the dam.

Three Gorges is the world's largest hydroelectric dam, holding back a reservoir 300 miles long that will before long contain 32 million acre-feet of water. Unleashing that water downstream would stand a good chance of creating the worst manmade disaster in history. It could dwarf the blasting of the Huayuankou dyke, which released the full fury of the Yellow River in northern China back in 1938. Even a credible threat of such an action would cause panic among the 350 million people downstream on the Yangtze. If the reservoir were emptied to neutralize such a threat, China's war machine would be crippled by the loss of electricity.

When the antidam campaigner Dai Qing interviewed scientists opposed to the construction of the Three Gorges Dam back in the 1980s, one military expert told her, "War is the key fact that determines whether or not we should construct the Three Gorges project. Are we going to make a sword of Damocles that will hang over the heads of future generations for decades to come?" They did.

VII

When
the rivers
run dry...

civilizations fall

21

Elisha's Spring and the Mysteries of Angkor

The first known permanent human settlement was on the west bank of the Jordan River. Jericho was constructed some nine thousand years ago as the ice age faded and *Homo sapiens* began to prosper in the warmer postglacial world. The settlement was modest enough. It covered barely 10 acres and had a thick defensive wall. Inside were a few hundred people and a stone tower, which survives today as reputedly the world's oldest manmade structure. Close by was a spring, recorded in the Bible as Elisha's Spring. It was the reason for Jericho's existence. As the Book of Kings puts it, "The hand of the Lord came upon him. And he said, Make this valley full of ditches. For thus saith the Lord, Ye shall not see wind; neither shall ye see rain, yet that valley shall be filled with water." The spring gushed into ditches that distributed its water to the fields and orchards.

The original town was destroyed by floods in the valley about eight thousand years ago. Jericho has been rebuilt and destroyed several times since, but the spring and the farming that it sustained just kept going. The area around was vital to the development of modern farming. It was probably the first place where people cultivated wild grains. By six thousand years ago, the inhabitants were growing peas and beans and olives and vines and figs. Soon the hills were carved into terraces that captured their own water from the rains. Today the spring still delivers water, at a regular 20 gallons a second, into a small pool, as if nine millennia of human history had never happened. And the farms still grow crops in what must be one of the longest-lasting and most durable agricultural systems anywhere in the world.

During the thousands of years since the fields of Jericho first flourished, many grander civilizations have come and gone in the Middle East, many of them based on apparently more sophisticated methods of catching and manipulating water. These are the places where Western civilization is deemed to have begun. It was around 7500 years ago that Sumerians in the "fertile crescent" of what is now Iraq constructed the first large irrigation systems using river water. They diverted water from the Tigris and Euphrates rivers down long canals and erected earth defenses against the spring floods. And they began to build great cities, too, like Ur and Kish and Uruk, where the first writing was produced and the first sciences developed. Uruk eventually had a population of 50,000.

These early Sumerian cities fought the first water wars. Some five thousand years ago, the king of Umma cut the banks on canals that neighboring Girsu had dug to the Euphrates. The breaches spilled water across the plain and destroyed Girsu's ability to feed itself. But the kings of Girsu were not beaten. They dug a new canal to tap the waters of the Euphrates' twin river, the Tigris. That gave them a larger empire than their rivals, and they ultimately saw the end of Umma.

But the Sumerian fields gradually became blighted. The bumper wheat harvests began to fail. Wheat gave way to barley. Then the barley too waned. And with that the fields became barren, the civilization foundered, and the land returned to desert. When the British archaeologist Sir Leonard Woolley excavated Ur in the 1930s, he wondered at the contrast between the great civilization he was uncovering and the barren land around him: "To those who have seen the Mesopotamian desert, the evocation of the ancient world seems well-nigh incredible, so complete is the contrast between past and present. Why, if Ur was an empire's capital, if Sumer was once a vast granary, has the population dwindled to nothing, the very soil lost its virtue?"

Woolley's successors believe they have solved the riddle. The problem seems to have been tiny amounts of salt coming down the river and accumulating over the centuries in the fields, eventually poisoning soils and crops. Cuneiform tablets of 3800 years ago describe a farm system in its death throes, recording "black fields becoming white" and "plants choked with salt." And that would explain why wheat, which is less tolerant of salt than barley, gave way first. Manipulation of the rivers made the Sumerian cities, but ultimately it destroyed them too. Salt chased civilization through Mesopotamia as mercilessly as any barbarian horde.

While the fields of Sumer turned to dust, new cities grew up farther north, in the lands around modern Baghdad. These societies culminated around A.D. 500 in the Persian Empire, which constructed canals to carry water from the Tigris to irrigate fields for 200 miles on either side of the river. This remaking of the landscape far outstripped the efforts of the Sumerians. Indeed, nothing bigger has been built in this region until today. It was "a whole new conception of irrigation, which undertook bodily to reshape the physical environment," said the American archaeologist Thorkild Jacobsen. In places the canals were 300 feet wide, and according to Sir William Willcocks, a leading Victorian engineer, "must have been capable of quite crippling the Tigris." The Persians in Iraq may thus have been the first to make a great river run dry.

The Persians had learned to keep salt at bay by avoiding overirrigation and planting weeds during the fallow season to keep the water table low. But they faced a new problem when silt washed into the irrigation systems. They solved it by employing thousands of slaves to dredge the waterways. But when Islamic rulers took over from the Persians, the dredging was left undone, and at some time in the twelfth century silt overwhelmed the system. Willcocks, who spent several years dreaming of reconstructing the canals, wrote of a "terrible catastrophe which in a few months turned one of the most populous regions of the earth into desert."

The lessons for modern times in the threats posed by salt and silt are clear. But for historians there is something else here that is, on the face of it, a little strange. Why cities? It is not obvious that rich agrarian societies should need them at all, and yet from Mesopotamia to the Nile in Egypt, and from the Indus in Pakistan to the Yellow River in China, a series of great civilizations grew up by irrigating arid areas from mighty rivers—and they all had cities.

One school of thought holds that cities were largely born because these agrarian societies needed new kinds of social organizations to collect, distribute, and contain water on a large scale. They had to hire farmers or coerce slaves into digging and maintaining dykes and canals and watching for floods, and they needed to develop scientific skills like astronomy and mathematics to predict nature's whims. It was the American historian Karl Wittfogel who coined the phrase "hydraulic civilizations" to describe societies that are organized primarily around the need to manage water. He argued that societies often required cities to do this. "It is the combination of hydraulic agriculture, a hydraulic government, and a single-centered society that constitutes the institutional essence of hydraulic civilization," he said.

Not all academics agree. But it remains remarkable how many great early civilizations emerged in environments where management of water was the first priority. Wittfogel contended that until the industrial revolution, the majority of human beings lived within the orbit of hydraulic civilizations. Ancient Egypt developed its water-management skills so well that it maintained population densities along the Nile that were double those of modern France. In the Americas at one time, three quarters of the population lived in a few small centers of hydraulic civilization in Mexico and Peru.

Much of Europe did not need such sophisticated water management, because it could rely on rainfall to irrigate crops. Nonetheless, it was those arid areas that did have to manage water that innovated and developed in other ways, too. In the Middle Ages, Córdoba, the capital of Moorish Spain, sustained a population of more than a million people through irrigated agriculture, at a time when the largest city north of the Alps was London, with a mere 35,000 people.

It was not just the desert civilizations of the ancient world that depended on sophisticated management of water. One of the most stupendous pre-industrial civilizations was centered on Angkor Wat in Cambodia. The Khmer civilization rose around A.D. 800, reached its height under Hindu kings in the twelfth and thirteenth centuries, and crashed in the early fifteenth century. Its influence spread throughout Southeast Asia, and its great capital, one of the true wonders of the world, was set in the jungle at the head of the Great Lake on the Tonle Sap, the reversible tributary of the Mekong.

Angkor is known today for its array of temples, but in the past five years research has revealed that the Khmer were about much more than temples. They too were remaking their landscape, and water was at the heart of their works. The temples that we see today were simply the ceremonial heart of a huge suburban landscape that, at its height, was "by far the most extensive preindustrial city in the world," says Roland Fletcher, an Australian archaeologist working there. "It was like modern-day Los Angeles."

Satellite pictures show what is not so visible from the ground: huge networks of canals spreading out from the temples, linking reservoirs and rice paddies, ports and suburbs. Homing in on areas identified from the air, archaeologists are digging up the remains of communities engaged in mining and boat-building, in weaving and manufacturing salt. And all this, it is increasingly clear, was dependent on sophisticated water management.

Visitors to the Angkor temples can still see huge moats and reservoirs dotted among the temples. The largest reservoir, the West Baray, is 5 miles across and still holds water. Locals go swimming and take boats out to a small temple on an island in the middle. Others reservoirs, such as the East Baray, are now empty and encroached on by jungle. And yet, as huge as they were, these barays were only baubles. The new evidence is that the workaday job of keeping the hundreds of thousands of Angkor inhabitants and their fields watered was done by waterways connected to suburban pools and areas of rice paddy that grew three and sometimes four crops a year. It was this that underpinned the wealth and the splendor of the rulers and their temples.

The Khmer civilization, arguably the greatest that Southeast Asia has ever seen, eventually faltered. What went wrong? The latest evidence points to an environmental catastrophe brought on by a failure of the water supply. Matti Kummu, a Finnish hydrologist who is interpreting the hydrology for an international team of archaeologists working in Angkor, says that the barays and ponds and canals and paddy fields all drew their water from streams draining from the surrounding hills into the Great Lake. The great urban center was established beside the lake precisely to use those waters.

The channels that collected the water and delivered it to the suburban complex were beautifully built and sometimes as much as 130 feet wide. But they were also artificial channels and prone to break out of their allotted course. They required constant maintenance. Kummu says that over the centuries one of them, the Siem Reap channel, started to take more and more water, and as it did, it began to cut its bed ever lower. In the end, the channel became so low that it could no longer deliver water to the barays and the ponds. The distribution network was left high and dry. (The West Baray has water in it today, Kummu points out, only because it now gets it from another source.)

I found the Siem Reap channel in among the temples. It is still flowing and has indeed eroded a deep gorge some way from its original course, where several piers of an old stone bridge stand forlornly in the bush. Locals have rather elegantly installed a large waterwheel that scoops water out of the bottom of the channel and pours it into a pipe at the top to irrigate a small tree nursery. Farther on, where the channel skirts the East Baray, it is 40 feet below the surrounding land and evidently could not fill the reservoir.

There are many unanswered questions about the collapse of the Khmer civilization. But it is increasingly clear that just as management of water created this quite mind-boggling civilization in the jungle, so too must failures of management have destroyed it.

And maybe that is what happened in the jungles of Central America, where the ruins of the Mayan civilization, like those of Angkor, today emerge from regrown rainforest. Here, great pyramids loom amid the trees as a reminder of a culture that, starting some three thousand years ago, cleared large areas of the forest and converted it into fields and cities and suburban areas that became one of the most densely populated regions on the planet. The Mayans created an urban society that established universities for mathematics and astronomy, pioneered the cultivation of corn, and built large reservoirs, viaducts, and canals to grow crops all year round.

Reservoirs at Tikal, a Mayan city in northern Guatemala, could hold enough water to keep 10,000 people for up to eighteen months. At Coba, the Mayans built dykes to increase the capacity of a natural lake. The central Petén region, the civilization's jungle heartland, may have had a population of more than 10 million. But after two millennia of success, the Mayan civilization suddenly and mysteriously collapsed in the ninth century. The collapse was so complete that when the Spanish arrived in the region a few hundred years later, just 30,000 people, less than 1 percent of the former population, were left roaming the jungle. And life was so poor that even expert looters and pillagers like Hernán Cortés and his army almost died of starvation there.

What went wrong? The most likely theory is that changing climate overtook the Mayans. Tree rings and lake cores reveal three catastrophic droughts in the final century of the civilization, which peaked in 810, 860, and 910. The lakes all but dried out. Presumably the reservoirs suffered similarly. Perhaps there were brutal wars for what water remained. Perhaps disease and starvation overtook the parched cities. At any rate, a civilization that prospered for two millennia by taking advantage of the fecundity of the rainforest and its water supplies eventually succumbed when the climate turned against it and the rivers and reservoirs dried up.

What can we make of all these tales of collapse? Two things. First, that civilizations built on intensive use of water often find themselves highly vulnerable to climate change or to insidious and destructive elements in their own systems of water exploitation, like salt and silt or the eroding power of artificial water channels. And second, perhaps, that less intensive and less grand uses of water—such as those employed around Jericho—can be more flexible in the face of change and so longer-lasting. Jericho never grew as big as the famous hydraulic civilizations around the Middle East. But while those civilizations fell long ago, farmers still make a living in Jericho. That too may offer lessons for our management of water today.

22

Losing the West

The big question is this: how vulnerable are our modern societies to hydrological breakdown? When the rivers run dry, will our own civilizations buckle? Is salt waiting to turn our best endeavors to dust? One place to look is the American West, where the epicenter of the coming water crisis may be Phoenix, Arizona.

With a population now well past 3 million, the metropolis of Phoenix is one of the fastest-growing urban areas in the United States. It sprawls across the Sonoran Desert with no apparent limit. In the mid-1990s, it was believed that Phoenix was growing by more than two acres every hour. It is scarcely less now. Metropolitan Phoenix covers nearly 400 square miles so far—by chance, roughly the size of the old Angkor urban zone.

Phoenix is as profligate with water as it is with land. The residents of private estates in smart suburbs such as Scottsdale and Paradise Valley water their lawns and fill their swimming pools as if they lived anywhere but a desert. Private lakes are blossoming. Developers compete to offer the greenest golf courses, the most luxuriant gardens, and the tallest fountains. In Fountain Hills, one of the outer suburbs, they have installed the world's highest-shooting fountain: it spurts 600 feet into the air for thirty minutes in every hour throughout the day. In such ways, Phoenix contrives to be among the biggest urban water users on the planet. Through the summer, residences in some suburbs consume around 1000 gallons of water a day.

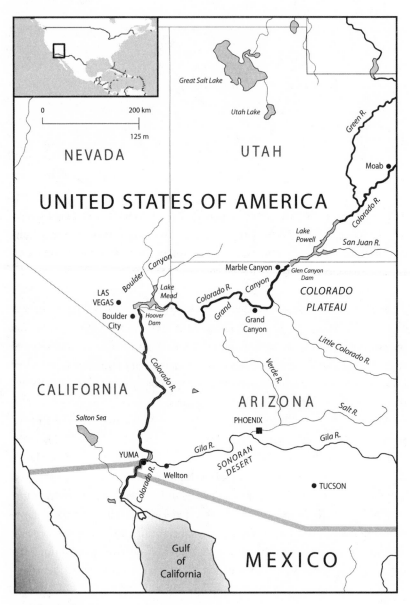

Colorado River

There used to be a river flowing through Phoenix—the Salt River. In pre-Columbian times, Hohokam Indians used it to irrigate fields. But dams have long since dried up the river down here. The Phoenix red-light district was built across its old bed. Phoenix and its suburbs grew increasingly reliant on water pumped from beneath the desert. But water tables have been falling fast. Arizona now pumps up more than twice as much as the sparse rains can replenish. Even projects in Scottsdale and elsewhere for recycling water back underground have not been able to make up the difference.

The end of Phoenix's profligate water use would have come by now if it had not been for the city fathers, who as early as the 1950s began lobbying Washington for cash to bring in water from out of state. Like California, Arizona saw its future in tapping the great waterway of the American West, the Colorado River. After twenty-five years of persuasion and the installation of an Arizona senator as chairman of the Senate Committee on Appropriations, pork-barrel politics delivered the goods, and in the 1980s the Bureau of Reclamation built for Arizona one of the world's largest and most expensive water-delivery systems.

The Central Arizona Project takes almost 1.6 million acre-feet of water a year out of the Colorado River and pours it into a concrete canal 300 miles long that zigzags across the desert to Phoenix and its smaller sister, Tucson. The project cost $3.6 billion to build and another fortune to run, and it loses 7 percent of the flow en route to evaporation. In recent times, the canal has been taking more than a fifth of the entire flow of the Colorado—much more in dry years. Its hydrological bounty unleashed the latest real estate boom in Phoenix, but it could be the final straw for the beleaguered Colorado. And it could take down the American West with it.

———

It is a truism that water won the West. The 1450-mile Colorado, which drains a twelfth of the continental United States, is the lifeblood of seven states, delivering its water to burgeoning cities, feeding irrigation projects, and generating hydroelectricity. Since the 1930s, many of its beautiful canyons have been flooded to make reservoirs. So much water is captured that the amount that makes it to the sea has fallen to nearly zero, leaving the Colorado delta to shrivel in the sun. A once rich landscape where jaguars and beavers roamed, it has not seen fresh river water since 1993. And no river water means no silt to maintain the delta, which is growing ever more vulnerable to tidal erosion.

Nature has been sacrificed to the demands of Uncle Sam's farmers. But now the river itself is faltering. And from the snow-covered mountains of Wyoming and Colorado to the desert cities of California and Arizona, the beneficiaries of the Colorado are getting worried.

Two giant reservoirs control the flow of the middle reaches of the Colorado and insure supply to the lower states. The first, Lake Mead, was filled in the 1930s behind the Hoover Dam in Boulder Canyon. Then, in 1964, the Glen Canyon Dam drowned a series of spectacular gorges to create Lake Powell, which was named after John Wesley Powell, a one-armed Civil War hero who in 1869 made the first boat journey by a white man down the river.

The two reservoirs collect water when snowmelt in the Rocky Mountains fills the river and distribute it to cities and fields during the long summer growing season. Having more than four times the capacity of the river's average annual flow, they can also even out fluctuations between wet and dry years.

While urban areas are taking an increasing amount, most of the water abstracted from the river still goes for irrigating some 4 million acres of fields in the river valley and in Arizona and California. America has always subsidized farming in the West, and today perhaps $1 billion a year is poured into keeping farmers irrigating crops that they would not otherwise grow. And subsidies encourage waste. Every year several million acre-feet of water evaporate from reservoirs, farm ponds, and flooded fields, while much of what does get to crop roots is used to grow low-value crops such as alfalfa. Even in dry years, the presumption is that a wet year will be along soon. But that presumption looks increasingly foolhardy.

The Colorado is both legally and hydrologically one of the most regulated rivers in the world. But it is becoming clear that the legal and the hydrological no longer mesh. A century ago, more than 20 million acre-feet of water flowed unimpeded to the Gulf of California every year. When lawyers shared out the river's waters between the states in 1922, on the eve of the dam-building era, they gave 7.5 million acre-feet to the upper basin states of Colorado, Utah, Wyoming, and New Mexico and another 7.5 million acre-feet to the downstream states of California, Arizona, and Nevada. With another 1.5 million acre-feet assigned to Mexico, that added up to around 16.5 million acre-feet.

That should have left water to spare, but ever since, flows have been diminishing. Since the compact was signed, the average flow has been 13 million acre-feet. From 1999 to 2003, the average sank to 7 million acre-feet—worse by far even than the Dust Bowl years of the 1930s. In 2002, it fell to just 3 million acre-feet. The U.S. Geological Survey says the Colorado hasn't seen a drought like this in five hundred years. Whether the cause is cyclical or global warming is unclear. But the survey says it can see no end in sight. The wet years that refilled the reservoirs have simply vanished.

Lake Powell was full to the brim in 1999, but by the end of 2004, after several years of drawing down the reserves, it stood three fifths empty. Its 170-mile-long reservoir could, if the drought continues, be empty by 2007. Lake Mead is scarcely better off. It is, of course, quite possible that between the time this is being written and the time you are reading it, the drought will have broken. Wet years can be very wet. The reservoirs could be full and the crisis could be over. Better rains in early 2005 brought some relief, but the crisis continued and the Bureau of Reclamation said that at average river flows, it would take the reservoirs ten years to refill. If the drought persists, the whole region is in deep trouble. And even if it breaks, most climate forecasts suggest that it will soon return with even greater force.

So in 2005 the states were starting to prepare for the worst. Since there was no more water, they deployed lawyers instead. The lower states, their attorneys contend, have a guaranteed entitlement, regardless of the state of the river: their share has to be provided from the reservoirs. If Lake Powell falls much lower, then the upper states will have to give up their share, emptying their own small reservoirs and halting abstractions from the river to meet downstream commitments. The upstream states, understandably, disagree. Colorado is not happy at being required to export three quarters of the snowmelt from its mountains when its own farms and cities are running low. Denver and Colorado Springs don't quite see why they have to shut off their sprinklers so Phoenix, Las Vegas, and Los Angeles can keep theirs on.

The upper states say they may contest certain downstream uses. That would include the Central Arizona Project. They argue that as the last user to come online, the canal to Phoenix and Tucson should be the first to give up access to the river in a crisis. In late 2004, Interior Secretary Gale Norton appeared to agree with them. As one Denver newspaper put it, with what sounded almost like relish, "It could stack up as the biggest water war in the West. Arizona could get shut off completely." Because there has never been such a crisis before, the differing interpretations of Western water law have never been contested. They soon may be. It could happen in 2006. But at the end of the day, if there is no water, there is no water. As campaigners from WWF say, it is now "not a matter of if but when the Colorado River plumbing system collapses."

Environmentalists meanwhile chose the moment of maximum decline in Lake Powell to resume a long campaign for the Glen Canyon Dam to be dismantled in order to "bring back the old canyon to its former glory," as the founder of the campaigning Glen Canyon Institute, Richard Ingebretsen, put it. The drought, he says, shows that the reservoir serves so little purpose it could be removed. Others say it shows why the reservoir is needed, to catch water in the infrequent wet years. With the reservoir increasingly empty, it seems a rather academic argument. As Ingebretsen said, either in hope or in expectation, "It's gone. It will never fill again."

———

Most observers of the hydropolitics of the Colorado believe that the days when most of the waters of the mighty river could be used to irrigate crops cannot last much longer. Cities are demanding a larger share of the pie. And the size of the pie is declining. But if water shortages don't put the farmers out of business, salt may be the apocalypse awaiting the great American agrarian civilization. Salt could kill off parts of the United States as certainly as it did Ur and Urek. As Arthur Pillsbury, the doyen of water resources in the American West, said to me before his death, "The Colorado basin will eventually become salt-encrusted and barren because of salt." The only question is when.

All down the river, more and more salt is clogging up the system. It is flowing downstream from the headwaters in the Rockies. It is also being washed from soils and bedrock in irrigated areas like Paradox Valley in Colorado and Wellton-Mohawk in Arizona. The river and the extensive manmade irrigation and drainage networks that circulate its waters have also become a vast system for collecting and distributing salt. Each year about 10 million tons of salt enter the system, but virtually none reaches the ocean.

Almost all the water flowing down the Colorado leaves the river several times to irrigate fields and returns via drains. At each step it both loses volume, through evaporation, and picks up salt. So the concentration of salt in the water increases as it travels downstream. At its headwaters, the Colorado contains about fifty parts per million of salt. By the time it reaches the last dam, near Las Vegas, it contains more than 700 parts per million. Tens of millions of dollars are spent on farms every year trying to minimize the problem, but even so, annual crop losses from salt are currently estimated at $330 million.

Outside Yuma, just upstream of the border with Mexico, is the ultimate technical solution to the salt problem. There lies a giant, $300 million desalination plant. It was built more than a decade ago to clean up the briny flow from the Wellton-Mohawk Irrigation District, the last in the United States, as it returns to the river close to the border. The idea was to deliver the cleaned-up flow to Mexico as part of the U.S. treaty obligation to supply 1.5 million acre-feet of usable water annually for irrigation over the border.

The Yuma plant has never been used. The desalted water would cost ten times more than its value to farmers. Instead, through the wet 1990s, when there was water to spare, the United States supplied Mexico with water from the main river. It sent the untreated Wellton-Mohawk brine down a canal to the Colorado delta, where it is the only water the delta gets. But as the reservoirs upstream empty, the desalination plant may be dusted down and prepared for use.

———

Phoenix is so called because it rose from the ashes of the Hohokam civilization—a society that probably ran the most extensive irrigation system north of Mexico. Why did it die? We don't know for sure, but it looks like salt brought it down. Of course, whatever happens, the American West won't be destroyed in quite the same way as Mesopotamia or the farming culture of the Hohokam. Even if its lifeline to the Colorado is cut, Phoenix won't disappear as completely as Ur or Urek. But the lessons from the past are potent nonetheless. Rivers can run dry. Salt can tighten its grip. Great cities can fall as a result. And even if the cities keep going in the American West, desert farming does seem doomed in the many places where it currently relies on the water of the Colorado.

23

Aral Sea: The End of the World

About 3 miles out to sea, I spotted a fox. It wasn't swimming. The sea as marked on the map is no longer a sea. The fox was jogging through endless tamarisk on the bed of what was once the world's fourth largest inland body of water. In the past forty years, most of the Aral Sea in Central Asia has turned into a huge uncharted desert. For the most part, no human has ever set foot there. This new desert is adding dry land twenty times the size of Manhattan every year. It cannot be long before someone decides that it should be protected as a unique, virgin desert. At present, though, such is the scale of what has happened here that the UN calls the disappearance of the Aral Sea the greatest environmental disaster of the twentieth century.

Until the 1960s, the Aral Sea covered an area the size of Belgium and the Netherlands combined and contained more than 800 million acre-feet of water. It was renowned in the Soviet Union for its blue waters, plentiful fish, stunning beaches, and bustling fishing ports. Most atlases still show a single chunk of blue. But the new reality is very different. The sea is broken into three hypersaline pools containing only about a tenth as much water as before. The beach resorts and promenades lie abandoned. The fish died long ago. As the fox and I peered north from near the former southern port of Muynak, there was no sea for 100 miles. It felt like the end of the world.

What has caused this environmental Armageddon? The answer lies in the death of the two great rivers that once drained a huge swath of Central Asia into the Aral Sea. The biggest is the Amu Darya. Once named the Oxus, it was as big as the Nile. In the fourth century, Alexander the Great fought battles on its waters as he headed for Samarkand and the creation of the world's largest military empire. It still crashes out of the Hindu Kush in Afghanistan. But like its smaller twin, the Syr Darya, from the Tian Shan Mountains, it is today largely lost in the desert lands between the mountains and the sea.

During the twentieth century, these two rivers were part of the Soviet Union. And Soviet engineers contrived to divert almost all their flow—around 90 million acre-feet a year—to irrigate cotton fields that they planted in the desert. This was one of the greatest ever assaults on major rivers. Perhaps nowhere else on earth shows so vividly what can happen when rivers run dry.

This chapter records my journey to discover how this disaster happened, to talk to its victims, and to find out what can be done now that Moscow's rule has ended. It was a deeply depressing trip. I found a landscape of poison, disease, and death in what was once one of the most prized areas of the Soviet empire. I found mismanagement of water on an almost unimaginable scale—a scale that has turned a showcase for socialism into a blighted land. More disturbing still, I found that in the aftermath of the collapse of the Soviet Union, nobody seems to have the vision or the will to rethink how this land and its rivers might serve the people living here better.

After Moscow went home, the governments of the region did set up an International Fund for Saving the Aral Sea and stated their desire for the sea to return. Soon afterward, I went to the region and heard them speak about their hopes at a big conference. But when I returned in 2004, the situation was worse than before. Even less water was reaching the sea than in Soviet times. In fact, most years now, no water at all makes it all the way down the Amu Darya, and the Syr Darya has less than a quarter of its former flow. And that is deliberate policy. I discovered that there is no plan to save the Aral Sea. Whatever the public statements, the engineers today, as in Soviet times, regard water flowing into the sea as wasted. Their policy is to keep the water back and allow the sea to dry out forever.

———

First, a little history. Central Asia has a long tradition of using the waters of its two great rivers to grow crops. In the days when Alexander the Great and the

Mongol conqueror Tamerlane invaded these lands, when cities like Samarkand and Bukhara flourished on the Great Silk Road, the people used the land and water carefully. Much of the region was covered in orchards, vineyards, and grain fields. Then the Russians came. The czars in the nineteenth century first saw the potential for planting cotton in the desert. They realized that the combination of near constant summer sun and water from the great rivers could produce cotton harvests to rival those of the United States. But it was the Bolsheviks who got down to business, turning ambition into action with socialist fervor.

Lenin lectured the southern republics of the new Soviet Union in 1921 that "irrigation will do more than anything else to revive the area, bury the past, and make the transition to socialism more certain." Under his successor, Stalin, the region's farms were turned into Moscow-run collectives, growing cotton for the textile mills of European Russia. "Commissar Cotton" had arrived. An ever-growing network of irrigation canals supplied water to billions of cotton bushes planted each spring on millions of acres of fields. Nations of nomads and cowboys and orchard tenders were turned into a near-slave society of cotton pickers. Dissent was ruthlessly crushed. "You cannot eat cotton," the prime minister of Uzbekistan complained in 1938. He was swiftly executed for "bourgeois nationalism."

By 1960, after several decades of obedience to Commissar Cotton, the canals were removing a staggering 32 million acre-feet of water from the rivers. But the Aral Sea remained full, partly because the rains had been good and partly because the irrigation systems eventually returned much of the water to the sea as drainage. But Moscow demanded more. Between 1965 and 1980, the area of irrigated land more than doubled. Central Asia became one of the largest irrigated areas on the planet, with some 20 million acres of fields crisscrossed by canals that could stretch to the moon three times over. By its own lights, all this was a dramatic Soviet success story: central planning at its finest, with everyone yoked to the common purpose of clothing the Soviet empire.

By the 1980s, 85 percent of all the fields in the Aral Sea basin were growing cotton. Orchards and vineyards, vegetable patches and wheat fields, even sports fields were given up. Almost every citizen was drafted to pick cotton. They began picking in searing summer heat that could reach 120 degrees

Fahrenheit, and ended in November when frost froze the fingers. Prisons and mental asylums and schools were emptied, factories and offices were shut, and herds of animals were abandoned for the duration. Buses were halted by police and the passengers were allowed to continue their journeys only after they had stripped a field of its cotton. Nobody was excused: not nursing mothers, not students, not doctors, not their patients. Only government officials remained at their desks, counting cotton.

But this marvel of organization carried the seeds of its own destruction. The newest canals were delivering water to the driest areas with the poorest soils. Farmers poured over 6 feet of water onto these fields each year, whereas the earlier fields required just over 3 feet. And increasing amounts of water never returned to the river in drainage. It accumulated in waterlogged soils, evaporated from fields, or drained into the desert, where new lakes formed. The Aral Sea had not disappeared so much as been broken up and dispersed.

Much the biggest reason for this dispersion was the creation in the early 1960s of the Karakum Canal, which tapped the Amu Darya as it poured out of the mountains and took much of its flow west for some 800 miles across the desert of Turkmenistan. Turkmenistan was the driest, emptiest, least populated republic in the Soviet empire. The canal was the longest and biggest irrigation canal in the world, and it turned Turkmenistan into the empire's biggest and most profligate water user. And it was the death knell for the Aral Sea.

In its first forty years, the canal has taken nearly 400 million acre-feet of water from the Amu Darya. Crucially, it has taken that water right out of the basin of the Aral Sea. None of it ever returns, even as drainage water. It was after the completion of the Karakum Canal that the sea began to empty in earnest. By 1990, when the Soviet empire imploded, it was receiving a tenth of its natural flow. Its volume was down by two thirds.

Many have called the emptying of the Aral Sea a classic Soviet blunder. The Russian geographer Grigori Reznichenko made a marathon expedition across the lands of the sea in 1988 to help expose the tragedy, and said afterward, "Everything happened as it would in a fairy tale, where everybody acts without being aware of the final result. In the end, tens of millions of tons of cotton outweighed and overwhelmed the Aral Sea." But the truth is more chilling. The draining of the sea was in fact perfectly deliberate. In a museum at Nukus on the Amu Darya delta, I found a series of maps drawn by Soviet engineers in the 1970s depicting the planned demise of the sea. By 2000 they expected it to be almost empty. Economic development plans once decreed that the seabed itself should be converted to cotton production as it dried up.

——

Today Commissar Cotton is gone. But his legacy remains, and as I discovered, his engineers are still in business. Since 1990, the new governments that rule the Aral Sea basin have increased the area of land under cultivation by a further 12 percent, and water abstractions from the rivers are up by a similar amount, to roughly 65 million acre-feet a year. That is considerably more than the 50 million acre-feet that, after evaporation, the rivers themselves once provided to the Aral Sea. The difference is made up by the drainage water that does return to the rivers, contaminated with salt and farm chemicals. But these days not even drainage water gets to the sea.

Some things have changed down on the farm. A market economy has emerged for crops like wheat, rice, and sunflowers. But cotton remains easily the region's biggest export. Today in Uzbekistan, the biggest producer, the government is still the only purchaser, and meeting cotton production targets remains a national obsession. During the harvest season, cotton employs a staggering 40 percent of Uzbekistan's workforce, including hundreds of thousands of schoolchildren. Every province, every canal network, and every farm has its production target. Even as the old collective farms are privatized, the targets persist, and farmers and officials can lose their land and jobs for failing to meet them. And cotton still consumes most of the region's water.

In October 2004, during my visit, the government declared that Uzbek cotton production had exceeded 3 million tons for the first time in several years. Ministers brimming with pride were interviewed for TV while standing in cotton fields. Officials who had seemed uptight and nervous suddenly relaxed. The bottles of vodka came out. Nobody cared that in the process, the ratchet on the Aral Sea had been given one more turn.

The amount of water used here is simply insane. Today the countries around the Aral Sea—Uzbekistan, Kazakhstan, Turkmenistan, Tajikistan, and Kyrgyzstan—occupy five of the top seven places in the world league table of per capita water users. Turkmenistan and Uzbekistan, the two countries that take their water from the Amu Darya, use more water per head of population than any others on earth. The Aral Sea basin is very far from being short of water. The problem is the simply staggering level of water use.

———

I traveled through Uzbekistan, the heartland of the old Soviet cotton empire, from its capital, Tashkent, in the far east, along the old Silk Road through Samarkand and Bukhara, and then north following the Amu Darya through desert toward its delta and the final, fateful destiny with the bed of the Aral Sea. What struck me was not the Soviets' single-minded determination to convert water into cotton but the sheer chaos that had resulted.

I drove first through the arid Hungry Steppe, west of Tashkent, where around 2.5 million acres are cultivated on one giant treeless cotton farm. All along the road, broken water channels leaked their water into the ground, and a salty white residue on the top of the newly harvested fields revealed massive overirrigation and poor drainage. Maintenance of the irrigation network has largely broken down since the Russians left, said my guide, Iskandar Abdulaev, from the Tashkent office of the International Water Management Institute. Cotton yields have halved in the salty soil. The dreams of the sixties are over. "The farmlands here are coming back to where they once were," he said. "They are returning to desert."

Beyond Samarkand, we passed through Bukhara. Its ancient mosques are a major tourist attraction. But it has also been a center of irrigated agriculture for four thousand years. Abdulaev grew up here. As a child, he picked cotton every summer, and he later wrote his doctoral dissertation on the state of the irrigation system. As we sped past more waterlogged and salt-encrusted fields, he explained the extent of the hydrological chaos here. The fields were for a long time irrigated from the local river, the Zarafshan, which drained into the Amu Darya. But by the 1960s so much water was taken from the Zarafshan that it dried up for the last 125 miles, so Soviet engineers dug a new canal and pumped water back uphill from the Amu Darya to Bukhara.

Yet still the water is wasted. Some 40 percent of all the water pumped uphill to these farms spills into the ground, collects in ponds, or seeps into the desert. Waterlogging has brought the salts in the soils to the surface, where they form a toxic crust after each harvest. Half the fields of Uzbekistan are salinized; in Bukhara the figure rises to 80 percent. The only practical way of removing the salt is to apply yet more water each spring to wash the salt into the drainage ditches that surround every farm.

So farmers are stuck on a treadmill, applying more water to grow their crops, which poisons the soils with ever more salt, which can be removed only with yet more water. In many areas more water is now used for flushing salt from soils before the cotton is planted each spring than for irrigation itself. But still the salt accumulates. Now the drains can't cope, so huge areas of land are being abandoned. Cotton yields are crashing on the fields that struggle on, so incomes are falling, too. Nothing you see here is new. Everything is old Soviet-issue. What cars are on the road are mostly old Ladas from the Soviet era. The world that Commissar Cotton created is crumbling.

We drove on, following the Amu Darya through the empty desert. Watering this landscape proved too much even for Stalin's engineers. But somewhere off to the west, the Karakum Canal was all but emptying the river into neighboring Turkmenistan. My first sight of the Amu Darya showed that it still contained some water, but little of it was clean mountain water. Most of it was by now briny drainage water returned to the river from places like Bukhara. Then, over a ridge of sand dunes, I saw a giant reservoir: the Tuyamuyun reservoir, built in the 1980s, which catches this briny wastewater and diverts it one last time to keep in business the fields of the last two provinces on the river, Khorazm and Karakalpakstan. The most obvious thing about this reservoir is how shallow it is: a few yards deep at best. The water is smeared across the desert, evaporating in the sun. Another piece of chaos.

And so to the Amu Darya delta. Politically, the delta is the autonomous republic of Karakalpakstan. Until the arrival of Commissar Cotton, the Karakalpaks were a nomadic cattle-herding and fishing race, whose name means the "black hat people." Many of them still wear black hats, though usually peaked caps made of cheap Russian leather. The Soviets and Uzbeks have treated them with disdain, using the republic for developing biological weapons and housing the notorious Jaslyq concentration camp. Even the ubiquitous security roadblocks give out here. Nobody, it seems, can imagine these beaten people causing trouble. This, their land, is the end of the line: the last stop before the Aral Sea. Its population of 1.5 million occupies the cockpit of a human crisis that is afflicting the region in the wake of the disappearance of the sea.

Here I met officials in overcoats and black hats, sitting in drafty offices, their cold fingers leafing through huge ledgers as they attempted to manage the ramshackle irrigation systems and divide up the salty drainage water delivered to them by their upstream neighbors. I met some who had trained in Moscow, even the United States. They need never have come home to this brackish backwater, but they did, out of patriotism to Karakalpakstan. They were trying to make the best of a very bad lot indeed, and were proud to take me on a tour of their crumbling empire.

Things were not good. Each year there are fewer tractors and more donkeys. The giant tractor-parts factory at Kanlikol, with its train tracks waiting to bring in fresh raw materials from across the Soviet Union, has long since closed. One telling statistic: diesel sales in Karakalpakstan have halved since the Russians went home. Children as young as eleven years old are still shipped out for months of cotton picking, working ten hours a day for a dollar or so to bring in the harvest on which national salvation apparently depends—though it increasingly seemed to me that cotton was the curse rather than the savior of the nation.

While most of the Soviet-style central planning has been kept in place, Soviet cash is not available, and the Soviet-built system is in a state of disrepair. Canals leak. Sluice gates are broken. Drainage ditches are silted up and filled with weeds. Fields have not been not leveled, so water ponds up uselessly in depressions. Many corrugated fields are wasting 30 percent of their water as a result. Some 60 percent of water intended for farms here, in the most water-short region of Uzbekistan, does not reach the fields.

One senior official told me, "Twenty years ago Karakalpakstan produced more than a million tons of crops a year—cotton, rice, wheat, and vegetables —but now we produce less than 300,000 tons." The precious waters of the Amu Darya are being used to less and less effect. Allowing for losses along the canals, farmers now take between 4.9 and 6.6 acre-feet of water to irrigate a single acre of land—equivalent to flooding every field to a depth of over 6 feet. Half of that water is needed to leach the previous year's accumulation of salt from the soil. Even with this profligate use of water, farmers in many areas get just one ton of cotton, a fraction of the old crop. Two fifths of all the fields of Karakalpakstan—half a million acres—have fallen out of use since the 1980s, partly from want of water and partly because the soils are clogged with salt. Half of that loss has been since 1999.

We drove past miles of abandoned land and empty farmhouses. The state farms have mostly gone bankrupt, and even the government's privatization program, which essentially gives the land away, has failed to tempt many. Why take on land that will never grow anything again? Tamarisk scrub now thrives amid the salt left behind by the last irrigation. Much of the land looks indistinguishable from the old seabed. It is returning to desert. That evening, the local TV news proudly announced that Karakalpakstan had met its cotton target for the first time since 1993. What the anchorman forgot to mention was that the government target had been halved since 1993. Failure was dressed up as success.

Poverty is spreading like a disease. There are fearful memories of the drought that overtook the republic in 2001 and 2002. Karakalpakstan got less than 30 percent of its water entitlement, and parts got no water at all, not even drainage water. Relief agencies put out famine alerts around the world, but the world was preoccupied with Afghanistan and Iraq, and nobody took much notice.

These days, even drinking water is hard to come by—this in a country with the second highest per capita water use in the world. On one former state farm west of the republic's capital, Nukus, people drink from the briny, chemical-laden water scooped from the irrigation canals. When the drought dried up the canals in 2001, they did not have even that. "I had to walk 2 miles to get drinking water for my family," one old farmer told me when we met in a chilly outbuilding.

On my visit, two years after the end of the drought, I saw women and children lining up with buckets at a hand pump, one of five thousand installed by the World Bank at the height of the drought. But they were fearful. Two thirds of the pumps had gone dry. It turned out that they had mostly captured water percolating from irrigated fields. As the fields have been abandoned, the pumps have had nothing to tap.

All this is a staggering development for a people who once thought they were part of the socialist revolution. All the billions of dollars and trillions of rubles invested over decades across this landscape have left many people little better off than the inhabitants of rural Africa. This is not a disaster that grabs you as you walk down the streets of Nukus or Muynak or the countless fly-blown towns and villages across the delta. The people are too familiar, in their pullovers and shabby black jackets and flat caps. And there are no refugee camps or soup kitchens. But probe a little and you find a blight that cuts right through a society that once prided itself on its order and ability to provide for all. And it affects everyone.

———

Nobody can escape the deteriorating climate. Once, the Aral Sea moderated the harsh desert environment here—cooling summers, warming winters, and insuring rainfall. Since it disappeared over the horizon, the summers in Karakalpakstan have become shorter and three degrees (Celsius) hotter, the winters colder and longer. Every farmer I spoke to said that he used to plant his cotton in March, but now it was May or even June. "It's not warm enough to grow cotton anymore," one told me.

Rainfall has declined, too, and the region is increasingly ravaged by dust storms. An estimated 77 million tons of dust from the seabed blows across the landscape every year, carrying with it an alarming cocktail of farm chemicals taken to the sea in past decades in drainage water. They include long-lasting pesticides like lindane, DDT, and phosalone. Traces have been found in the blood of Antarctic penguins and in Norwegian forests. But most of the poisoned dust falls close to the former sea, deposited on fields, inside houses, and, worst, down the lungs of children. An average year now has fifty days of dust storms.

Alarming though the spread of pesticides on the wind may be, most researchers believe that there is a still more devastating threat in the dust storms —salt. The stuff is everywhere in Karakalpakstan. There is no escape. It comes on the wind, down the irrigation canals, and through the pipes carrying drinking water from reservoirs; it is in the vegetables the Karakalpaks grow in their gardens and the fish and birds they catch out on the delta; it is left behind on the soil surface by the irrigation process. Salt destroys the perilous productivity of the land, uses up precious water in flushing it out of soils, creates poverty, and ultimately kills the people themselves. Salt is the true tragedy of this land. Worse than the poverty, worse than the water shortages, worse than the pesticides, the land and its people are being poisoned by salt.

Sixty percent of the rural population of Karakalpakstan drink from wells dug into the salty underground reserves in the desert. In many villages the water is so salty that milk added to tea instantly curdles. Mothers in some villages say their children will not take their breast milk because it contains too much salt. And salt is a relentless, silent killer.

I talked for a long time to Oral Ataniyazova, a gynecologist and campaigner for the health of her people. Ataniyazova grew up on the delta, where her father was first secretary of the local Communist Party. International acclaim for her work here has included the Goldman Environmental Prize awarded in 2000. She could have fled this blighted land but decided to return to work full-time for her people. Salt, she said, has turned Karakalpakstan into a nation of anemics.

Among the 700,000 women in the tiny republic, 97 percent suffer from anemia, five times the rate in the early 1980s, three times that elsewhere in Uzbekistan, and probably the highest rate in the world. "All of our women are sick—and so are all our newborns," Ataniyazova said. Anemia causes a disturbing number of pregnant women to die from hemorrhages during pregnancy and childbirth, and 87 percent of their babies are born anemic.

Aigul Aspanova, at the maternity unit of the Kanlikol hospital, told me, "The children are slow and get sick a lot. We also have lots of birth defects, especially of the mouth, legs, and hands." One in every twenty babies is born with a defect. Infant mortality, at seventy-five children per thousand births, is the highest in the former Soviet Union. Karakalpakstan also has the highest rate of cancer of the esophagus in the world, among the highest rates of tuberculosis, and unusually high rates of other cancers, immunological disorders, kidney and liver diseases, allergies, and reproductive pathologies, said Ataniyazova. Half of all deaths here are caused by TB and pneumonia. Average life expectancy in these border regions around the Aral Sea has fallen in thirteen years from sixty-four to fifty-one.

"The entire population of Karakalpakstan has been chronically exposed to salt and farm chemicals for a long time," said Ataniyazova. "We have examined the literature but can find no research into a health situation like this anywhere in the world."

I talked about this later with a group of farmers meeting in Chumbai, one of the towns on the delta. These aging, weather-beaten men in shabby jackets, pullovers, and hats against the encroaching winter chill, told me that nobody escapes the salt. "We get salty water, put it onto salty land—and then the winds bring another layer of dust as well," they said. These men had lived tough lives. They were certainly not experts at diagnosing medical conditions, but when I asked if they saw any health effects among themselves and their families from this toxic environment, they all nodded.

My interpreter tried to shut off the conversation. It was too difficult for them to talk about, he said. But the manager from a state farm rose slowly to his feet, as if pronouncing at a funeral. He clutched his hat in front of him and looked me in the eye. "You can see it in the faces of everyone living here," he said. "We are all affected. There is a lack of good blood. All the women have it, and it is worse in pregnancy. Many children are born deformed here. In the drought years, lots of people died, especially the children. My daughter and son were both in the hospital for six months. Every family has someone like that."

There was more nodding, but nobody would say any more. This was private grief among old men with sick wives, sicker children and grandchildren, failing crops, growing poverty, and a poisoned land. Was I meeting the last farmers of Karakalpakstan? The last to live and die by the Soviet dream?

———

This delta certainly seems to be a place where things end. Back in 1972, somewhere near Chumbai, the last tiger in Central Asia—the final wild member of the Caspian tiger species in the world—was hunted down in one of the last forests on the delta. A few decades before, the last remnants of a rebellion against Soviet rule in Uzbekistan make their final stand here, finally meeting their end in a hail of bullets at a place called Blood Lake.

These days, it seems to be the entire landscape and entire population that is reaching the end. The 2700-square-mile delta was once a rich natural ecosystem, with dozens of lakes and woodlands full of wild boar, deer, and huge numbers of fish. As late as 1960, hunters took 650,000 muskrat pelts from the delta—a quarter of the muskrat fur production for the whole of the Soviet Union. But today the woodland has gone, most of the lakes have dried up, and there are fewer than a thousand muskrats left. And now the farming system that killed nature is in its turn being destroyed.

As you get closer to the Aral Sea, the proportion of abandoned fields increases, until you see that almost nothing grows. But as salt and water shortages have destroyed the fields here, the authorities plan to rehabilitate the delta by diverting the upstream drainage water that once irrigated the fields into a series of new lakes. The bulldozers and earth-moving machinery that once dug canals are now remaking the delta once more. In his office in Nukus, I met the smartly dressed Ashirbekov Ubbiniyaz. As a first secretary in the republic's government in the 1990s, he helped prop up the system that created the current crisis. As the local boss of the International Fund for Saving the Aral Sea, he was now in charge of the lake-making project. He dismissed my suggestion that the water should be allowed to flow on into the sea. "We can't recover the Aral Sea," Ubbiniyaz told me. "We need to use to the maximum the water we have from God."

You could see the superficial logic, and the locals were grateful. I met a fisherman, Mohammed, on his way to the largest of the new manmade lakes. "When there is water flowing into the lake, like this year, there are plenty of fish," he said. By some counts, two or three tons of fish emerge from the lakes on a good day now. The prices are pitiful—we paid less than three dollars for an 11-pound fish in the market at Muynak, and less than two dollars for a large, freshly shot pheasant. Heaven knows what they contain in the way of chemicals. But that is the least of the concerns of the locals. As one asked me, "What should we do? It's not just the fish. We are all polluted."

A few fish cannot hide the increasing poverty and suffering of the people of the area. The lakes can keep only a minimal population active. There is already an alarming rate of migration out of the delta. A fifth of the people have left, and that seems bound to accelerate. The delta could one day be as empty as the Aral Sea bed itself. But not necessarily.

People trying to find a way forward here say the first thing to do is forget both cotton and the old Soviet ways of servitude. Ataniyazova is appalled at the passivity of her people in the face of a crisis that threatens their very survival. "It's important to get people to speak up. If we keep silent, our mentality will kill us," she said. Ecologically, she noted, the real need is to retire irrigated land and revive the old ways of herding livestock and tending orchards and growing vegetables. "We don't need cotton, we need fruit and vegetables," she said. Then both the people and the sea could recover.

It is easy to forget that most of the people in Karakalpakstan were nomads and cattle herders before they became cotton pickers. They have always kept a few animals; it was one of the few areas of life that neither Moscow nor Tashkent wanted to interfere with. And now that the state has so comprehensively failed them, the Karakalpaks are building their herds again. The abandoned fields provide new pasture. Almost every family has a cow or two. Most people told me that if they got any extra cash, they would buy more cattle. An adult cow is money in the bank, fetching about $150 in the local markets. Morning and evening, the delta roads that once hummed with tractors going home from the fields are now full of cattle going home from grazing on the abandoned cotton fields. The old ways are coming back.

———

And so to Muynak. Back in the 1950s, when the sea still lapped the shores of Muynak, Soviet filmmakers portrayed the heroism of fishing fleets on the Aral Sea. The ships caught 48,000 tons of sturgeon, carp, and bream each year. Ferries sailed from Muynak to Aralsk, its companion port on the north shore in Kazakhstan. But there is more sand than sea between Muynak and Aralsk today. Muynak last saw the sea in 1968. The last trawler set sail in 1984. And the corroding remains of the trawlers stranded on the seabed in Muynak harbor have become one of the totemic images of the sea's demise.

Another is the crumbling fish-processing factory at Muynak. It was once the largest fish factory in the Soviet Union, employing thousands of workers and producing 6500 tons of fish products a day. It had its own power station. When I first came here, in 1995, the workers were still canning fish caught in the delta and, ludicrously, brought by train from the Baltic. It was a dark, Dickensian operation, with pasty-faced women in cheap scarves chopping and gutting fish on wooden benches and spraying the unwanted entrails across the floor.

This time, the same manager greeted me at the same grandiose factory gates. But all he could show me now was a dirty and dingy shed where, two or three days a month, a handful of workers clocked on to convert the smelly remains of a couple of tons of fish bought in the local market into dry, powdered fish meal for sale as cattle fodder. The rest of the factory stood derelict. The manager dreamed of spending half a billion dollars to revive the plant. He said the government had drawn up a plan. But this seemed pure delusion. Where would the fish come from? Where would the money come from? The sense of decay was as pungent as the lingering smell of fish.

Muynak still bravely keeps its fish insignia. The mayor's office has a map showing the old sea lapping the town's shores. Paintings of the heroes of the trawlers still cover the walls of shuttered offices in the fish-processing factory. It is a town living on memories. Once, the townsfolk remember, the balmy climate meant they could go swimming in the sea in November. Now they have their overcoats on before the end of October. The population back in the sixties was over 40,000; today, after a mass flight of what might be called ecological refugees, it is around 10,000. The sole income for many families is the pensions of their senior members.

Muynak seems like the end of the line. But beyond it is Uchsai, or "Tiger's Tail," a name it got because it sits at the very end of a peninsula that once jutted out into the Aral Sea. If Karakalpakstan is the end of the line for the Amu Darya, Uchsai is the end of the line for Karakalpakstan. Back in 1995, I found it one of the most depressing places I had ever visited. On my return it was even worse, like a ghost town out of the Wild West. Its population had fallen from 10,000 to 1000. And few of these ghostly, emaciated people stayed for any reason other than lethargy and lack of an alternative.

The town's one source of employment, a fish-smoking yard that limped on in 1995, had closed down two years later. Worse, on the day it shut, the piped water supply for the town was cut off. Nobody there now seems to know why. It just happened. A couple of days later, a water tanker began twice-weekly visits from Muynak. It is Uchsai's only source of water. Every family buys about 50 gallons a week—about what the World Health Organization says is the minimum water requirement for a day—at a cost of about 20 cents, which is all they can afford.

I met two elderly schoolteachers in the town's only street. "The sea left here in 1961," one of them remembered. "It came back in 1966 and 1968. But that was the last time we saw it." For a while the trawler crews, who included both the teacher's parents, followed the retreating sea. But as it disappeared over the horizon, they gave up. The last trawler returned, with empty nets, in 1981. Now they have only the sand and the salt. When I first came here, the workers at the fish-smoking plant told me that they expected the sea to come back one day.

"It has dried up before and returned," one explained. Not now. The women said it was gone forever.

A couple of boys played football in a listless way in the road. I asked them what they would do when they left school. "We will go to Kazakhstan to find work," they said without enthusiasm, before sidling off, totally uninterested in who we were or why I was there. "Most of our children are sick here," said one of the teachers. "They have anemia. You can tell. They are weak and slow and cannot study well." She blamed the sand and the salt in the air, the poverty and the lack of fresh fruit and vegetables. "We used to grow vegetables in our gardens when we had piped water," she said. "But we can't do that now."

There have been epidemics of TB and birth deformities in Uchsai. "During the drought many children died here," one of the teachers told me. Had they received any assistance from the outside world? One of them remembered that the World Bank took a look, but nothing came of it. I later discovered that none other than the president of the World Bank, James Wolfensohn, one of the most powerful people in the world, had touched down for a few minutes in Uchsai during a helicopter ride over the Aral Sea back in 1995. But they were right: nothing came of it.

You could be forgiven for thinking that the government in Tashkent had no interest in this forgotten corner either. But a third of a mile down the road, I spotted a gas flare. The state company Uzbek Gas has been drilling for gas beneath the Aral Sea bed and built a gas compression station right outside Uchsai. But its large, fenced-off complex brings no benefits for the people of the town. "They never stop, they don't talk to us, they fill our town with gas smells and give us no compensation. They don't even buy things in our shop," the teachers complained. "But their trucks are destroying our road. When the snow falls in winter, the water tanker won't be able to get here. Then what will we do?"

On the way back from the end of the world, as we bounced and swerved among the potholes along the isthmus from Uchsai, we drove past the notorious ship's graveyard, where rusting hulks are blasted by sandstorms. For many, this scene has become a symbol of the disappearing Aral Sea. But for me the symbol is now the betrayed people of Uchsai. Once almost surrounded by water, they now line up in the main street with old milk churns, awaiting the water tanker. When the river ran dry, the tide really went out for them.

VIII

When
the rivers
run dry...

we go looking for new water

24

Taking the Water to the People

It is the world's largest civil engineering project and aims to remake the natural hydrology of the world's most heavily populated nation. But it began rather inauspiciously, one morning in Beijing in April 2003, when the city's vice mayor, Niu Youcheng, accepted a bottle from a visiting local official. The bottle was filled with water from a reservoir on the Yangtze River 800 miles away in the south of the country. Its handover signaled the start of China's south-to-north transfer project, which is intended to be the ultimate solution to the desiccation of the Yellow River and the North China plain.

Some people see the scheme as an exercise in engineering hubris and a disaster in the making. But Chinese leaders say it is a logical extension of the grand schemes of great societies down the ages to remake their hydrology. Ancient Mesopotamia and Egypt both harnessed their rivers to feed their people. The Romans were famous for their aqueducts. Persian empires were built around laboriously excavated tunnels that delivered water from deep underground. But even at their greatest extent, these ancient works were mild modifications to natural drainage patterns. Today's engineers have bigger ambitions, diverting entire rivers onto distant plains. And China's scheme is the most ambitious ever to have got under way. It will, we are told, enable the Middle Kingdom to continue feeding itself as it has done for thousands of years. Maybe.

The south-to-north scheme will divert part of the flow of the Yangtze, the world's fourth biggest river, to replenish the dried-up Yellow River and the tens of millions of people in megacities that rely on it. The price tag is $60 billion, more than twice the cost of even Colonel Qaddafi's fantasy-world Great Manmade River Project. And it aims to deliver twenty times more water than Qaddafi's pipe dream.

Some of the water will come quite soon. In Beijing, people should be drinking Yangtze water regularly in 2007. Certainly it will be there in time to fill the swimming pools and pretty the streets with fountains as Beijing hosts the Olympic Games in 2008. Some of the water will take longer to get there, but within twenty years, say the planners, the project should annually be siphoning north three times as much water as England consumes in a year. The costs may be colossal, but China says the south-to-north project cannot be allowed to fail.

The project is actually three separate diversions. Two of them are already under construction. The first one will enlarge the existing Danjiangkou reservoir on the Han River, a major tributary of the Yangtze, and take its water north. The reservoir is already Asia's widest artificial expanse of water; enlargement will flood another 140 square miles and displace a quarter of a million more people. The canal north will be 200 feet wide and as long as France. As it crosses China's crowded plains, it will span 500 roads and 120 rail lines and tunnel beneath the Yellow River through a giant inverted siphon.

The second will take water from near the Yangtze's mouth across Shandong Province on the North China plain and deliver it to the megacity of Tianjin, which has suffered chronic water shortages since the 1990s. Part of it will use the 2500-year-old Grand Canal, which was the world's largest artificial river in preindustrial times and the first to have lock gates. Today it is a sump for effluent from China's burgeoning industry, but there are plans to clean it up.

The third, western route is the most ambitious. It alone is expected to cost $36 billion. It will take water from the Yangtze headwaters amid the glaciers of Tibet and push it through tunnels up to 65 miles long into the headwaters of the Yellow River. There is no firm route yet, but several tributaries will be tapped, and there is talk of building the world's highest dam. This route will be the only one to deliver water directly into the Yellow River. The other two are intended to relieve pressure on the river by supplying water for cities and farms on the North China plain that currently take 8 million acre-feet a year from the river. But altogether the plan is to send some 36 million acre-feet of water a year from the Yangtze to northern China.

China's leaders love huge projects. Modernism lives on in their souls. With the World Bank claiming that China has already lost $14 billion in industrial production from water shortages, the scheme seems to them like a sound investment. But even as the first earth was dug, fears were growing about escalating costs. And academics and water planners I met in Beijing in 2004 raised a range of concerns.

The middle route, they said, could cause an ecological crisis on the Han River, taking a third of its water and worsening an already serious pollution problem. Wuhan City, a busy river port with a population of 3 million, could become a cesspit overnight. What, they wondered, about the cost of relocating refugees from the Danjiangkou reservoir? Could the filthy and decrepit Grand Canal really be cleared of pollution? Is the engineering intended for Tibet more than a figment of someone's imagination? And since China is trying to move to more realistic pricing for water, won't the transferred water be far too expensive for the intended recipients to buy?

China has a vibrant antidam community, which honed its arguments over the Three Gorges project. It sees the scheme as another megalomaniacal folly and wants the billions to be spent instead on improving Chinese water efficiency. Ma Jun, a journalist and campaigner in the mold of Dai Qing, says, "Chinese factories use ten times more water than most developed countries to produce the same products. Chinese irrigation uses twice as much." Even old-fashioned Chinese toilets use much more water than their Western counterparts. The United States has been able to grow its economy for the past thirty years without increasing water use, he says, and so can China. And in the long run, he argues, it would be easier to shift the focus of Chinese food production from the northern plains to the south—where the water is.

But the antidam contingent lost the war over Three Gorges, and they seem set to lose this one too. Already China's scheme is being taken as a template for an even larger project in the world's second most populous nation, India. During 2003, as Chinese earthmovers embarked on their south-to-north project, politicians in Delhi were lining up to back plans for what is known as the River Interlinking Project. It would redraw the hydrological map of India in a quite breathtaking way by harnessing the great monsoon rivers of the north, like the Ganges and the Brahmaputra, and sending their water south and west to the parched lands where the droughts are worst and the underground water reserves are sinking fastest.

———

The Indian scheme is far more complex than the Chinese plan. It would build dozens of large dams and hundreds of miles of canals to link fourteen northern rivers that drain the Himalayas. And it would pump their waters south along 1000 miles of canals, aqueducts, and tunnels and through 300 reservoirs to fill a second network, linking the seventeen major rivers of the country's arid south. These rivers include the Godavari, the Krishna, and the Cauvery, each of which has been diminished by heavy overabstraction for irrigation.

The transfer would involve moving 38 million acre-feet a year, very similar in scale to China's scheme. But with so many more rivers and links, the price tag would be two to three times higher, with official estimates ranging from $112 billion to $200 billion, or around 40 percent of the country's GDP. While the Chinese project sounds doable, whatever the pitfalls, the Indian scheme sounds like a great leap into the unknown.

Even so, the River Interlinking Project did not come out of nowhere. It has a long history. The legendary British engineer Sir Arthur Cotton first conceived of something similar in the mid-nineteenth century as a means of improving the country's transportation by linking rivers with canals. His scheme was dashed when the British decided to build railroads instead. But the fantasy never quite died. For a while it was poetically described as providing India with a "garland of canals." But it took a drought across India in the summer of 2002 to bring the idea back to the fore.

The specter of famine returning to India after more than a decade in which the country has had food to spare was a profound shock. Southern India, like northern China, seems to be on the verge of a hydrological crisis. And with water tables tumbling and the country's population predicted to increase by 50 percent within the next half-century, overtaking China to reach a staggering 1.5 billion people, there is a sense that a hydrological holocaust might not be far away. Surely, say many Indian politicians, the only answer is to replumb the nation.

In 2003, the president and prime minister, most state governments, and the Supreme Court—agencies that in India's complex democratic system spend much of their time in internecine warfare—united to promote the River Interlinking Project. Government scientists said it could provide enough water to increase irrigated farmland by more than 50 percent and to power 34,000 megawatts of hydroelectric capacity. On closer inspection, though, the blueprints show that as much as a third of this power would be needed for pumping water around the scheme's network of canals and tunnels.

Will it happen? There are plenty of objections. The rampant pollution in most northern rivers would result in a grand interchange of sewage down the canals. One estimate holds that an area of land the size of Cyprus would have to be flooded for the various structures, leaving 3 million people homeless. Even engineers in the states that apparently stand to gain the most are skeptical. Karnataka, one of the driest, is also one of the highest—often more than 2000 feet above sea level. Engineers there doubt that the water from the north would ever reach them.

A former head of the national water ministry, Ramaswarmy Iyer, calls the whole idea "technological hubris." India already has half-completed water projects that have cost billions of dollars, which should be finished first. He says that across arid India, three quarters of the monsoon rains still evade dams and wash into the sea. Better to try to collect that rain as it falls onto fields than to try to carry water across the country. "The answer is to harvest our rain better," he says.

Bangladesh objects too. It sits downstream of India on both the Ganges and the Brahmaputra. Bangladesh may fear the rivers' monsoon floods in summer, but it also relies on those rivers to irrigate its crops and recharge underground aquifers. It complained angrily in 1974 when India built the Farakka barrage on the Ganges close to the border, diverting valuable dry-season water flows into Indian irrigation canals. It blames the barrage for dried-up fields, disease, and a salty invasion of seawater into the Sunderbans mangrove swamps in the Ganges delta. In 1996, India promised not to reduce the flow further. Yet now it is talking about doing precisely that.

Like China, India is in a bullish mood at present, with annual economic growth at around 8 percent. Its engineers have always thought big, and now its politicians want to turn their dreams into reality. So it could happen. It is at least possible that within two decades, the two most populous countries in the world—with the two fastest-growing economies—will have spent a quarter of a trillion dollars on redirecting their rivers and remaking their hydrology.

———

And the urge to build big seems to be spreading. Spain spent 2003 lobbying the European Union to foot most of the $10 billion bill for its own north-to-south project. It wanted to build a 600-mile canal from the Ebro, the biggest river in the wet north of the country, taking as much as three quarters of its water to Murcia and Almería in the country's increasingly arid south. There, it was earmarked to irrigate more than a million acres of desert where spaghetti westerns were once filmed and to water golf courses at new tourist resorts.

The project bit the dust late in 2004, when a general election brought in a new government that had listened to critics of the plan, like Pedro Arrojo-Aguda, an economist from the University of Zaragoza. "It's clearly crazy," he told me just before the plug was pulled. For one thing, he said, it would wreck the Ebro delta. The 12-mile tongue of sandbanks, lagoons, and reed beds protruding into the Mediterranean is one of southern Europe's most important wetland nature reserves as well as Spain's largest rice-growing area and an important fish nursery. This land of flamingos and paella is already starved of water and silt from the river by upstream dams. The sea is eroding it. The transfer would seal its death warrant, he said. And for what? The golf courses and tourist resorts might be willing to pay for the new water, but the farmers could never pay. "The water will cost more than a dollar for 265 gallons—twice the current cost of the desalination of seawater. By the time it is delivered, few farmers will want it," Arrojo-Aguda said.

For Spaniards like Arrojo-Aguda, the Ebro transfer seems like a return to the bad old days of General Franco, the Fascist dictator who gave Spain more dams than any other similar-sized country. Perhaps that is why, in the heady days of their campaign, angry villagers from valleys in the Pyrenees that would be flooded for new dams joined irate farmers from the Ebro delta, city environmentalists, and political activists in probably the first ever protest to go to Brussels demanding that the EU close its purse. The new government, sworn in in 2004, heard the protests, assessed the economics, and opted to build desalination plants instead. But the project still has powerful backers and is only one general election away from a revival. And the drought that spread across southern Spain in the summer of 2005, reputedly the worst in half a century, brought renewed calls for consideration of the transfer plan.

Europe has other such schemes in the pipeline. The Greek government has a plan to take water out of the country's largest west-flowing river, the Acheloos, and pump it through a tunnel to irrigate tobacco fields on the eastern plains. It would partly destroy the Messolonghi delta, one of the last surviving untouched deltas on the Mediterranean.

In Britain, government agencies have proposed using the eighteenth-century canal system, dug to ship industrial goods around before the age of the railroads, to create a national water grid able to ship water from the wet north and west to the drier south and east. And London's water engineers have long dreamed of tapping the Severn River as it runs out of Wales and through the West Country to pump its water over the Cotswold hills and into the headwaters of the Thames.

———

Back in the days of heroic Soviet engineering, there was a plan to tap great north-flowing rivers of Siberia like the Ob and the Yenisei and send their water south to refill the canals of the Central Asian cotton industry, and just possibly to help revive the Aral Sea. First proposed by the czars back in the late nineteenth century and revived by Stalin in the 1930s, the idea became a serious proposition in the early 1980s. Thousands of scientists and engineers were recruited to help bring it about. But in the end it fell foul of a mixture of romantic Russian nationalists who did not want to give up their rivers to Muslim states far to the south and the reformist pragmatism of Soviet leader Mikhail Gorbachev.

But you cannot keep a bad megaproject down. In 2003, Igor Zonn, the director of Soyuzvodproject, a Russian government agency in charge of water management and ecology, told me, "We are beginning to revise the old project plans. The old material has to be gathered from more than three hundred institutes." It had won vocal support from ambitious politicians like Moscow's mayor, Yuri Luzhkov, a possible successor to Vladimir Putin as Russian president. If for no other reason, that link makes it a project to be taken seriously again.

The proposal would be roughly equivalent to irrigating Mexico from the Great Lakes, or Spain from the Danube. It would drive a 650-foot-wide canal southward for some 1500 miles, from where the Ob and the Irtysh meet, to replenish the Amu Darya and Syr Darya rivers near the Aral Sea. The canal, according to blueprints shown to me by Russian scientists, would carry 22 million acre-feet of water a year. Though this is just 7 percent of the Ob's flow, it would be equivalent to half the natural flow of the two Central Asian rivers into the Aral Sea.

Down among the "stans," they have always loved the idea. "Although it seems ambitious, it appears to be the only tangible solution to the ecological and other problems caused by the drying of the Aral Sea," says Abdukhalil Razzakov, of the Tashkent State Economic University in Uzbekistan. And early in 2004, Luzhkov visited Kazakhstan to promote the plan. He said that this time around, there would be no handouts from Moscow. Central Asia would have to pay a realistic price for the water. But behind the scenes, Moscow is starting to see the scheme as one of the water projects that could rebuild its political and economic power in the region.

Madness? Certainly. But as Nikita Glazovsky, a leading Russian geographer and former deputy environment minister under Boris Yeltsin, told me when we met in Moscow in 2001, the region's engineers "still find it easier to divert rivers than to stop inefficient irrigation."

———

In North America, the parched Western states of America have their eyes on Canada's water. It is not very surprising. Canada has some of the largest rivers in the world. When the spring snowmelt gets going, around a tenth of all the water in all the rivers in the world is gushing through British Columbia and Yukon and the Northwest Territories into the Pacific and Arctic oceans. And the Great Lakes, on the border between the two countries, are one of the largest freshwater reserves on the planet. Meanwhile, only a few hundred miles to the south, the High Plains are parched, the reservoirs in the Colorado River sit two thirds empty, and California imposes restrictions on its lawn sprinklers.

In 2003, President George W. Bush raised hackles in Canada by calling for talks about U.S. private companies buying some of Canada's water to supply cities in the American West. It could become the ultimate fulfillment of the old maxim that "water flows uphill to money." Exactly how the water would be shifted economically was not made clear. Bush was less interested in the engineering than in establishing that NAFTA, the North American Free Trade Agreement, could cover the continent's water reserves. But for many north of the border, the issue is highly emotional.

The idea also has unpleasant echoes of vast schemes devised in past decades. One would have drained part of the Great Lakes into the Mississippi. Another, dreamed up back in the 1960s by Donald McCord Baker, an official in the Los Angeles water department, would have captured the waters of the Columbia River in the United States and then, in south-to-north order on the map of Canada, the Fraser, the Liard, the Skeena, and finally the mighty Yukon and Mackenzie rivers.

Its backers, who included Ralph Parsons, the head of one of the world's largest civil-engineering companies, envisaged building dams up to 1800 feet high in the canyons of British Columbia to funnel water into the Rocky Mountain Trench, a natural depression that would have been turned into a 680-million-acre-foot reservoir—that is, one with three times the total capacity of all the manmade reservoirs in the United States today. The project would, the blueprint proposed, carry south 114 million acre-feet of water a year—three times the capacity of China's entire south-to-north project.

That particular plan is dead and buried. But Canadian water officials say the pressure is growing to allow small water exports, especially from the Great Lakes. And they fear that once a precedent has been established, California and Arizona could come calling. Everything has its price under NAFTA. If the current drought in the American West persists, and if climate predictions that there is worse to come prove right, then who knows?

Africa already knows a bit about harnessing great rivers. The Nile today is as dammed as it could be. Its waters power Egypt and supply its irrigators. Barely a drop reaches the sea in most years. But the continent's other great river—the world's second largest river through the world's second largest rainforest—is a new target. The Congo River runs undiminished throughout the year and has ten times the flow of other major sub-Saharan rivers, like the Zambezi. The veteran British colonial water engineer Henry Oliver, who built the Kariba

Dam on the Zambezi, called it "one of the greatest single natural sources of hydroelectric power in the world." He would love to have got to grips with it, he told me after he retired to South Africa.

And now his post-apartheid successors say the great river through what Joseph Conrad called the "heart of darkness" could soon be lighting up Africa. South Africa's state-owned utility, Eskom, is leading a five-country project to upgrade a small and largely moribund hydroelectric plant on the river at Inga and turn it into a dynamo that runs a pan-African electricity grid.

At Inga, just downstream of the Congolese capital, Kinshasa, 34 acre-feet of water a second rush down a series of rapids. Eskom's engineers say the force of this staggering volume of water at this point could generate 40,000 megawatts of power. That's twenty times more energy than comes from the Hoover Dam, ten times as much as Africa's current largest hydroelectric dam, the Aswan High Dam on the Nile, and more than twice that of China's Three Gorges Dam. It is "enough to light up Africa and export to Europe as well," according to the company's adviser on project developments, Ben Munanga.

The proposed $50 billion Grand Inga hydroelectric project would not even require a large dam, because unlike the Nile, the river is guaranteed to run strongly all year round. So there is no need to store water. Most of the money would be spent connecting this vast power source across Africa's jungle, bush, and desert to the continent's main population centers. The first power lines would link it to South Africa via Angola and Namibia, a distance of 1900 miles. Next, pylons would strut across 2500 miles of rainforest and swamp in the Central African Republic and Sudan to Egypt. Nigeria also wants to take Inga power to West Africa. Its power might even straddle the Sahara and enter Europe via Spain.

The plan is backed by South Africa's president, Thabo Mbeki, who hopes that it could become the flagship project of NEPAD, the New Partnership for Africa's Development, which is being promoted by the G8 governments in Europe and North America as a new Marshall Plan for the continent. It could just happen.

There is a surprisingly strong political push too behind another mega-project to harness the Congo. This time the idea is to send the water itself north to irrigate the fringes of the Sahara and refill Lake Chad. In 2002, several governments met in the rainforests of central Africa to sign an agreement on sharing the waters of the Congo. They said they wanted to raise $5 billion for a dam that would barricade one of the river's major arms, the Ubangi River, at a sleepy port called Palambo in the Central African Republic, and send the water north.

The Lake Chad Basin Commission, representing Nigeria, Niger, Chad, Cameroon, and the Central African Republic, fleshed out the plan in 2004. The dam should, they said, divert 80 million acre-feet of water a year over a narrow watershed into the headwaters of the north-flowing Chari River and down its wide floodplain before ending up 1500 miles away in Lake Chad. The survival of more than 20 million people in the lake basin depends on the Lake Chad Replenishment Project, the commission claimed. It could rehabilitate huge irrigation projects around the lake, such as the South Chad Irrigation Project, which were built in wetter times but are now unused.

"If nothing is done, the lake will disappear," claimed Adamou Namata, the water minister in Niger and the chairman of the commission. But the project would be "an opportunity to rebuild the ecosystem, rehabilitate the lake, reconstitute its biodiversity, and safeguard its people." In late 2004, the Nigerian president, Olusegun Obasanjo, agreed to pay $2.5 million for a feasibility study of the project.

On the Logone and Hadejia rivers, in Cameroon and Nigeria, poor farmers quake at the thought. For them, large engineering projects to remake the rivers have always brought trouble. Wherever the promised water went, it never reached them. Generally they lost water as it was corralled and privatized and sent uphill to money. They want and need very different solutions to their water problems, ones that go with the flow of nature rather than against it.

25

Sewage on Tap

I advise against eating vegetables bought in the markets of Vadodara, a large industrial city in southern Gujarat in India. Most likely they will have been irrigated with neat industrial effluent from the town's numerous chemical plants. The effluent pours down a drain that runs for 40 miles from the city's industrial zone to the sea. The drain is a river that "never runs dry," one farmer told me. "You can get as much water as you want, when you want it. It's an assured supply of water, unlike the wells here." (The wells, I need hardly mention, run dry because the chemical companies are emptying underground water reserves.)

The drain is not something you would want to go near. I peered beneath one of its concrete covers, which have everywhere been broken by farmers trying to get at the liquid. There was a tarry scum on the surface. It was jet black, though I was told that at different times it took different colors, depending on the effluent of the day. The stench was sometimes unbearable—a mixture of solvents and paints and much else that I couldn't recognize. But all along the road beside the drain, for dozens of miles, there were plastic pipes trained across the potholed surface, siphoning the toxic treacle onto fields and into villages. Legally, pumping from the drain was banned. But "during droughts especially, some of the officials of the company running the canal turn a blind eye," said my guide, a local investigator called Vaibhav Bhamoriva.

Dozens of firms, from the giant Hindustan Petrochemical Corporation to backyard outfits like the Shree Bitumen Company, dump their waste into the drain. Some 20 million gallons flow down it to the sea each day. And, said Bhamoriva, tens of thousands of people in more than forty villages rely on it to grow their crops. It is the basis for a vibrant local economy. After interviewing hundreds of farmers, he has calculated that the total output of crops from effluent-irrigated fields here is worth half a million dollars a year. Some farmers use the effluent to grow rice; more grow sugarcane or millet. Most frighteningly, a lot grow fruit and vegetables, which are sold in the city market. "The sellers mix the vegetables grown from effluent with all the other fruit and vegetables. So we all eat it," one farmer told me. "We farmers know about it, but in the town nobody knows what they are eating."

What is the health effect of all this? In truth, nobody knows. Officially, the farmers are not allowed to use the effluent, so officially there is no problem. I dropped in at Chokari, a village of 1500 people living on either side of the toxic canal. The villagers told me that they had no other source of water except a pump more than half a mile away that they use for drinking. The contents of the drain were used to irrigate fields of rice, millet, and vegetables. Farming with the stuff was not pleasant work. "Sometimes the smell when we are watering our fields is unbearable," one told me. "We avoid getting the liquid on our bodies, because sometimes it burns the skin. We watch the water, and if it changes color, sometimes we stop irrigating." They knew nothing about any longer-term effects on their bodies. But they said that it killed their soils. After three or four years, the chemicals destroyed the soil's fertility, leaving behind a cake of salt. They had to abandon the fields then. Eventually there would be no fields left, and perhaps no healthy farmers. But for now they made a living.

By a perverse irony, the drain in places runs less than a hundred yards from a canal built to carry clean water from the newly dammed Narmada River in the south of the state. But while the drain runs black and full, the canal has sat empty for some years. I spoke to one farmer who lost most of his land twenty years before to make way for the canal. He was resigned to the black humor of his situation. "Last year, during the election campaign, they released some water down the canal," he said. "But we've had none since." The toxic effluent, however, was always available.

Water should be recycled. Most of us would instinctively agree with that—

until, perhaps, we realize what that means sometimes. Should dirty water be recycled? Nobody should countenance using neat industrial effluent to water crops. But sewage may be another matter. After all, even in Europe the concentrated muck created at sewage treatment plants is sometimes packaged up and sold as fertilizer for farmers. And in high-tech Singapore, officials announced plans in 2003 to add volume to the city-state's main drinking-water reservoir by topping it up with a 2.5 percent dose of recycled sewage effluent. That is small potatoes compared with London, whose inhabitants drink water that has been drunk and excreted several times as it makes its way down the Thames, being extracted and returned by towns such as Swindon, Reading, and Maidenhead before it reaches the capital. The advanced water-treatment technologies in use at every step make it as safe as water anywhere, despite its unappetizing recent history. The water-treatment plants have become just another loop in the water cycle.

But across the poor, developing world, raw sewage is increasingly being used for irrigating crops. Chris Scott, of the International Water Management Institute, has conducted the first global survey of the practice. He has come to the staggering conclusion that perhaps a tenth of all the world's irrigated crops —everything from rice and wheat to lettuces, tomatoes, mangoes, and coconuts—are watered by the smelly, lumpy stuff coming out of the end of sewer pipes that empty the drains of big cities. Without it, much of the world would go hungry. In many countries—India, China, and Pakistan, to name just three of the biggest—there is very little sewage treatment, and yet a great deal of the sewage ends up being poured onto fields anyway, complete with disease-causing pathogens and sometimes laced with toxic waste from industry.

Scott estimates that more than 50 million acres of the world's farms are irrigated with sewage. And business is booming. The practice is most frequent on the fringes of the developing world's great cities, where clean water can be in desperately short supply in the dry season while sewer pipes keep gushing their contents onto the nearest open land all year round. In Hyderabad, the Indian city where he works, "pretty much 100 percent of the crops grown around the city rely on sewage. There is no other water available."

And however much consumers may squirm, farmers like it that way—first because the sewage is rich in nitrates and phosphates that fertilize their crops free of charge, and second because the supply is often much more reliable than clean water from rivers or irrigation canals, which means farmers can grow high-value crops that need constant watering, such as vegetables.

For these reasons, farms hooked up to sewage pipes make bigger profits than their rivals who rely on clean irrigation water. In Pakistan, farmers using sewage for irrigation typically earn $300 to $600 more annually than those without the benefit of sewage, says Scott. In West Africa, he met one farmer who grew twelve crops of lettuce a year on his sewage farm. You can see the benefits in land prices as well. In parts of Pakistan, it costs twice as much to buy fields watered by sewage pipes as neighboring fields irrigated with clean water. People downstream like the farmers, too. They are essentially operating a free municipal wastewater treatment service that stops rivers and reservoirs from stinking so much. And so, often secretly, do governments. The system feeds the people. In Pakistan, sewage irrigates a quarter of the country's vegetables.

Many would say that the risks outweigh the benefits. Those risks include disease among farmers and customers and environmental problems such as the buildup of heavy metals and unwanted nutrients in the soil and underground water reserves. "Right now," says Scott, "wastewater irrigation is in an institutional no man's land. Water, health, and agriculture ministries in many countries ban the practice but refuse to recognize that it is widespread." But, he says, instead of trying to outlaw it or pretending that it does not exist, governments ought to regulate it. "We need to recognize that sewage is a valuable resource that grows huge amounts of food. So instead we should help the millions of farmers involved to do it better."

That means keeping sewage effluent for irrigating nonfood crops or crops that will be processed or cooked before being eaten. Obviously, it is more dangerous to pour sewage onto a field of lettuce than a field of cotton, or even sugarcane. And ultimately it means working toward treatment to make sewage safe. In a handful of countries—most notably Israel, Jordan, Tunisia, and Mexico—that is already happening. Sewage is treated to remove pathogens before being released to farmers. In these countries, recycling forms part of a national strategy for maximizing water use and making sure valuable nutrients are not wasted.

Mexico recycles enough treated wastewater to irrigate around 600,000 acres. I visited a giant new state-of-the-art sewage treatment plant at Juarez, El Paso's twin city on the Rio Grande, which treats half the city's sewage and delivers enough effluent down a canal to irrigate 75,000 acres. It is virtually the only source of water for crops downstream on the Mexican side of the river. Israel converts around 70 percent of the wastewater from its cities into treated effluent for irrigating export crops such as tomatoes and oranges. This is an effective addition to its national water supply of about a fifth.

This makes good sense. In urban areas, almost every drop of water brought to the city to fill faucets eventually leaves again as sewage effluent or in industrial waste. Where it can be made safe, it should not be wasted.

26

Closed Basins and Closed Minds

The Salton Sea just happened. It was all the fault of Charles Rockwood, a land speculator in California's boom years at the start of the twentieth century. He and his buddy George Chaffey, who had already made a fortune planting orange groves in Los Angeles, dreamed of turning a desert depression close to the Mexican border into an agricultural boom town. They planned to do it by capturing some of the flow of the mighty Colorado River from Yuma, Arizona, sixty-odd miles to the east.

Despite the distance, it didn't look too difficult. The bottom of the depression they had earmarked for irrigation was 230 feet below sea level, the second lowest point in the whole United States. So once the Colorado bank had been breached, the water would flow downhill most of the way. And for part of this distance there was an old riverbed to channel the flow. In 1901, Rockwood's California Development Company constructed a rickety wooden dam and canal to divert some of the Colorado's flow.

Rockwood and Chaffey did a swift bit of rebranding. Their desert depression had come at a rock-bottom price partly because it was known to locals as the "valley of death." It sounded much better when renamed the Imperial Valley. The soils were good, too, and settlers eager to make their fortune came thick and fast. Within four years, some 14,000 of them had staked out farms, dug irrigation canals, and planted 300,000 acres of fields irrigated with Colorado water. Traders moved in to service them. New towns such as Calexico, El Centro, and Brawley sprouted. It was like a gold rush.

It seemed too good to be true, and it was, for with the water came silt. The Colorado is the second siltiest river on earth, after the Yellow River, and by 1904 its muddy waters had clogged Rockwood's canal. The water dried up, and crops were dying in the Imperial Valley. Facing down angry mobs of farmers, Rockwood cut a deal with a landowner over the border in Mexico and dug another canal to Imperial Valley down an old creek known as the Alamo. But disaster struck again the following year. The Colorado was in full spate when it broke its banks near the new canal and began pouring into the desert. The floodwater soon found Rockwood's canal, and by the end of autumn the entire flow of the Colorado—32 acre-feet a second—was coursing into the Imperial Valley. There, having nowhere else to go, it created an inland sea.

As the flood subsided, the river stuck with its new course. It seemed to have permanently given up its old route to the Pacific Ocean and settled on a one-way ticket into the desert. It set about remaking the region's physical geography to fit its needs. There was a small waterfall where the new Colorado entered the new lake. It soon began to cut back upstream, eating up the soft bed of the new river. Such was the power of the river that for a while the waterfall was retreating through the desert at a rate of more than half a mile a day. It gouged out a gorge 100 feet deep and 1000 feet wide. It was as if a new Grand Canyon were being created before the world's eyes.

The federal government panicked. Farms were disappearing beneath the sea, and nobody was quite sure where the water would go when the river had filled the desert depression. With millions of acre-feet of water pouring into the valley, there wasn't much time. So the government called in the Southern Pacific Railroad Company, which had already lost a chunk of its line from Los Angeles to Tucson. The company spent $3 million over eighteen months dumping 6000 railcar loads of rock, gravel, and clay into the desert in repeated efforts to force the Colorado back onto its old course. Finally, on February 10, 1907, it succeeded. But while the Colorado flowed once again on its old course south into Mexico and finally to the ocean through the Gulf of California, it left behind a new sea in the Imperial Valley covering some 600 square miles.

If all the farmers of the Imperial Valley had departed, the lake would probably have evaporated to nothing by now. But the farmers were not done. They regrouped and lobbied for a more permanent canal to bring Colorado water to California. The All-American Canal, finally completed in 1938, was one of the first and largest major diversions of water from the Colorado. It was finished soon after the Hoover Dam, which helped regulate the wilder flows of the river and prevent future disasters. The canal has since become the basis for a billion-dollar business in the Imperial Valley, providing salad crops for sale across the country. The descendants and corporate successors of the farmers who stuck it out through the bad times have grown rich.

And the sea is still there too, thanks to the large amounts of drainage water pouring from the farmers' fields. In its new role as a desert sump, it gradually became more polluted and saltier as pesticides and the salt brought down by the Colorado accumulated there. It soon became known as the Salton Sea. And the federal authorities had the idea of filling it with fish. First they tried striped bass, then salmon, halibut, bonefish, anchovies, and turbot. All died. In the 1950s they tried thirty species of fish from the Gulf of California, and some of these, especially orangemouth corvina, did much better.

Today, despite the accumulating toxins and occasional epidemics of disease among the fish, the sea is one of the world's most productive lakes, home to an estimated 200 million fish. And the fish attract birds in large numbers. Strange to say, with the rest of California draining its wetlands for new real estate, the toxic sump has become the state's premier bird habitat, providing a home for 380 species, including egrets, cormorants, brown pelicans, and various boobies. It is home to more species than the Florida Everglades, America's most famous wetland.

And it hasn't just attracted birds. In the sixties, before Las Vegas emerged as a desert oasis for the jet set, the Salton Sea was the place to be seen. Frank Sinatra came with his rat pack, and yacht clubs, beauty contests, and nightclubs followed. Even the Beach Boys put in an appearance on a shorefront that had everything except surf. Then nature intervened again. Heavy rains one winter flooded the yacht club and hotels, and the jet set took off for new pastures.

Water is a precious commodity in California, however. The Imperial Valley's farmers are legally entitled to three quarters of the state's share of the Colorado. An annual dose of some 3 million acre-feet of water into the valley insured that they continued to prosper. But change is afoot. For years California has been taking more than its legal entitlement of the Colorado's flow—5 million acre-feet a year, rather than the allowed 4.4 million acre-feet. And now, after years of drought on the river and with other states growing increasingly thirsty, Washington has ruled that enough is enough. On New Year's Day 2003, it began cutting flows to the Sunshine State.

To avoid having the cities run dry, the state decided to cut its needs by persuading the Imperial Valley farmers to give up almost a fifth of their share and let it roll on down the All-American Canal to San Diego. The state plans to make things right with the farmers by lining the canals that distribute water through the valley and investing in more efficient ways of irrigating, such as drip irrigation. Reluctant farmers finally accepted the deal in 2003.

It seemed like a neat solution. But there is a problem. If the farmers no longer flood their land, much less drainage water will flow off the fields and into the Salton Sea. With inputs expected to fall by at least a quarter, the legacy of Rockwood's rocky engineering a century ago will start to dry out. The director of the Salton Sea Authority, Tom Kirk, is understandably anxious to save his charge. He says his lake may be a sump, but it is also "a rich and vital ecosystem of international significance." He fears that with water that once filled the Salton Sea heading off instead to supply San Diego, his sump will empty, and within a decade all the fish could be dead.

"On-farm conservation," as the irrigation engineers call it, would trigger "off-farm Armageddon." Rather than funding the farms' increased "efficiency," Kirk would like the state either to divert other drainage water to the sea or to start shutting down farming so that more water is left for the sea. But the farmers, now sold on the idea of trading their water rights, are talking about new sales—to Las Vegas.

Will Kirk get his way? Much may hang on how California chooses to classify its toxic oasis. Is it, as the Rockwood saga suggests, a manmade lake? If so, it has no right to protection under California's conservation laws. But if it were decreed to be a natural lake, the law would be very different. It should be entitled to its own allocation of water to guarantee it survives.

Surprisingly, scientifically at least, the sea should have a case. It turns out that its existence is no accident. Large bodies of water have occupied the "valley of death" before, and they too got their water from the Colorado. Probably four times in the past 1500 years, the Colorado has changed course and poured its waters down the Alamo and other local creeks into the Salton depression. Each time the wayward river soon choked on its own silt and resumed its old course. But often it left behind so much water that the lake was for a while bigger than the present Salton Sea. There are tide marks on mountains around the edge of the valley to prove it.

Maybe during the 1905 floods, the Colorado would have rushed into the valley without the help of Rockwood's rickety canal. We will never know for sure. But if so, then the sea is, in part at least, natural—and entitled to legal protection. It is now a century since the Salton Sea formed. Maybe now, before it starts to die again, would be a good moment to decide the matter.

———

The story of the Salton Sea may be somewhat singular, born of reckless engineering during reckless times. But it tells an important story of increasing relevance in river basins around the world. More and more rivers are becoming like the Colorado: rivers in which every last drop of water is utilized, sometimes several times over. In hydrologists' jargon, these rivers that no longer, or only barely, reach the sea are known as closed basins. The only water left for nature is often drainage from humanity's last use. And every effort to utilize the water better—whether building dams or lining canals or improving irrigation efficiency or even harvesting the rain—can have unintended consequences for downstream users.

François Molle, of the International Water Management Institute, tells the story of a small river basin in central Iran. The Zayandeh Rud basin covers an area the size of Wales, where some 3 million people live by irrigated farming. Until modern engineers arrived, they relied on a mixture of wells, old underground tunnels that collect water, known as *qanats,* and traditional diversions from rivers called *maadi.* These eventually took most of the water, allowing a small amount to drain to a swamp, the Gavkhouni, where the last of the river died. But farmers wanted more water. So engineers brought in a new tier of modern water structures: large dams and canals. They promised to provide more water. But, says Molle, "Basically, the new dams and canals redistribute the water that was once supplied by traditional *qanats* and river diversions." Some people got more water, but only at the expense of others.

Undeterred, the government tried another tack. It put money into lining irrigation canals to cut water demand. Saving water always sounds good. But as a result, less water percolated down through the soils, water tables fell, and wells emptied. The government responded with bigger electric motors to pump water from deeper underground. At that point the *qanats* dried up.

Of course, there were winners and losers in this long process to grab every drop of water in the Zayandeh Rud basin. Bigger, politically better-connected farmers won. Small farmers and the age-old water distribution arrangements lost. The *mirabs,* traditional democratically elected water chiefs, lost their control and authority. But there was not more water. In fact, there was less. The growing volume of water held behind large dams increased evaporation. The endeavor was "costly and worse than useless," says Molle. "Dams and concrete disrupted what had been a nice, functioning system."

And that water, sitting in large reservoirs, set the stage for the next step— taking water out of the basin altogether. In the next valley, which contains the ancient Persian city Yazd, new boreholes have started to lower water tables and destroy the *qanats* that traditionally supplied the city. So the new plan is to take water there from Zayandeh Rud. The lesson, says Molle, is that in closed basins, "any change in water use or abstraction is likely to rob Peter to pay Paul. Or, in Iran, it will rob Yadullah's water to irrigate Said's garden."

———

There are dozens of similar examples around the world of how efforts to save water can rebound. We have looked at misguided efficiency measures in the Rio Conchos in northern Mexico. Intended to provide more water for U.S. farmers downstream on the Rio Grande, they are likely to lower water tables and deprive Mexican farmers of the underground water on which they increasingly depend. WWF's Hector Arias had it right there: the old idea that rivers are from Mars and groundwaters from Venus is dangerously misguided.

Yet water engineers, agriculturalists, and even environmentalists have proved slow to make these connections. They continue to promote the idea that if a hundred individual farmers "save" so much water by pouring less of it onto their fields, then the entire valley or river basin will have "saved" that figure times one hundred. But it isn't always so, because in many river basins, most of the "wasted" water would actually have moved on through the water cycle, either returning to rivers, from where someone downstream would capture it, or percolating underground, from where the same farmer or his neighbor might later pump it back to the surface.

Saving water can often be good, of course. Drip-irrigation systems and canal lining and the rest can all bring local benefits. And insofar as they reduce evaporation, they can provide victimless gains. But in closed basins, planners have to be very careful that water savings don't rebound to the disadvantage of others. One man's waste of water is often another man's source of water. The biggest danger is when farmers announce that some new water-saving technology will allow them to irrigate more fields or sell the water for use elsewhere. This may not be so much water saving as water theft. It may simply be depriving someone downstream of the use of their "wasted" water.

At the global scale, we need to be very careful not to conclude too easily that saving water, however locally beneficial, can solve the world's water problems. It will undoubtedly help, but only if seen as part of the wider water cycle. We can never separate one part of the water cycle for our attention while ignoring other parts. Water is the ultimate renewable resource. But that only makes the task of managing the water cycle prudently more critical.

27

Out of Thin Air

The dew hung heavy over the Sussex downs near the ancient Roman city of Chichester. Peering through the swirling mist of a November morning, I spied a pond. It appeared almost magical, like an oasis in the desert. Though the air was damp, this patch of water was far from any spring or stream, and too high to gather water from the surrounding pastures. And this was an unfailing pond. It had not emptied for decades, not even during the long drought of the mid-1970s, which had killed streams, dried up springs, and turned fields almost to desert.

I was looking at a dew pond. Its secret was to capture unseen moisture from the air in the hills. Though largely forgotten today, dew ponds were once essential to the great sheep pastures that covered these chalk hills. They are a hidden mystery of the English landscape, only now being recognized and revived, out of nostalgia and a desire to provide water for wildlife.

For centuries, dew ponds flourished across southern England. Gilbert White, a famous eighteenth-century chronicler of the English countryside, wrote of a dew pond high above his home in Selborne in Hampshire. It was, he wrote, "never above 3 feet deep in the middle, and not more than 30 feet in diameter, and contained perhaps not more than two or three hogsheads of water, yet it is never known to fail, though it affords drink to 300 or 400 sheep and for at least 20 head of large cattle besides." While the valleys dried up in summer, he said, the pond on the hilltop always kept its water.

Naturalists often imagine that dew ponds are prehistoric features. That is understandable: many of them still survive near ancient monuments. On the South Downs, I saw dew ponds near Iron Age forts at Cissbury Ring and Chanctonbury and close to the Celtic temples of Lancing Ring. In his poem "Puck of Pook's Hill," Rudyard Kipling talks of a time long ago "before the Flint Men made the dewpond under Chanctonbury Ring." And the celebrated war photographer Don McCullin hinted at the same antiquity when he opened and closed a book of his best photographs with ethereal shots of dew ponds close to Glastonbury Tor in Somerset, site of the first Christian church in England and reputed home of Camelot and King Arthur.

But most dew ponds are more recent. They were dug to provide water for the large flocks of sheep that grazed on the downs in the eighteenth and nineteenth centuries but abandoned when the decline in sheep and the arrival of piped water in the twentieth century rendered them largely obsolete. One of the last diggers was a Mr. Smith from the Chiltern Hills. He advertised his craft into the 1930s, providing a phone number for free estimates in "a secret process handed down from father to son over 250 years." His ponds were "guaranteed to condense and retain beautiful clear water without the aid of pumping etc."

So what was the "secret process"? On chalk hills, water quickly slips underground through the porous rock, so the first requirement is a waterproof lining. Dew-pond diggers used local clay, "puddled" to make it watertight by driving teams of oxen through it. But that doesn't explain where they got their water. There is a ticklish mystery here. Some water comes from rain, to be sure. But the dew ponds' real trick is to capture moisture from the air, catching the tiny water droplets in passing clouds and fog and encouraging more water vapor to condense on the ground in the form of dew.

Here, the hilltop location is the key, because as air rises it cools. Cold air can hold less moisture, so the rising air creates water droplets that form clouds or fog. And at night, as the hill air cools further, more water condenses and dew forms. Martin Snow, a local historian, has been mapping dew ponds on the Sussex downs for years. He says that many of them are cleverly sited at the heads of dry valleys that channel mists and moist air uphill. The old sheep farmers, he reasons, noticed that fog lingered at the heads of these valleys and the grass was often moist. "Creating ponds would have been a natural way to catch that water."

But the "secret process" guarded by the pond diggers was more than a matter of siting. Their trick was to create a pond that captured the maximum moisture by providing as cool a surface as possible on which dew would condense. They did this by putting straw beneath the clay and stones on top. The straw insulated the clay, keeping it colder than the soil beneath at night. The stones, which shed heat quickly at night, lowered the temperature further. Once a successful dew pond was created, it would, in effect, generate its own water from the air.

Many dew ponds have been plowed up in recent decades as pastures have given way to arable farming. Others are overgrown. On the hills above Worthing, residents still remember how shepherds would halt their flocks for a drink at the Tolmare Farm dew pond on the way to the Findon Sheep Fair. Today that pond is empty, with a radio tower through its center. At Cissbury Ring, military tanks have pierced the pond's clay crust. But a surprising number of these ponds just keep on catching moisture from the air, and dew ponds have benefited from recent national efforts to restore and dig ponds of all kinds.

The National Trust renovated one on Bignor Hill in an ancient landscape of Bronze Age barrows crossed by Neolithic causeways and a Roman road. The Sussex Downs Conservation Board dug a new one on the grounds of Sussex University in 2004. Sadly, says Snow, "Most of the renovators are more concerned with providing water for wildlife than with recreating dew ponds for their original purpose." The Bignor dew pond is full of reeds and yellow flag and the pink-flowered amphibious bistort. A worthy aim, of course. But it is fenced off from the sheep that it was dug to provide for.

Dew ponds do not, perhaps, stand comparison with other methods of catching moisture from the air, such as rainwater harvesting, which we explore in the next chapter. Their champions see them more as countryside curiosities than as a new paradigm of water management. But they do suggest that water is often to be found from unexpected sources. Dew collectors, for instance, turn up on several Croatian islands in the Adriatic. The otherwise waterless island of Vis has paved surfaces on hillsides behind a Franciscan church that collect moisture from hot, dry summer air and channel it into cisterns. On the Canary Island of Lanzarote, few vacationers will have noticed that local farmers surround their crops with a mulch of volcanic gravel that condenses atmospheric moisture.

Perhaps the most bizarre discovery was made a century ago in the Crimea by the Russian engineer Friedrich Zibold, who found thirteen pyramids of stones on hilltops in the forests around the ancient Byzantine town of Feodosia. Each was 30 feet tall and 100 feet across. They looked like burial markers. But Zibold noticed that they were surrounded by broken terracotta pipes that went to the city. The locals called the pyramids "fountains." He concluded that they were dew-collecting machines, and proved the point by building one close by that turned out 35 gallons of water a day for several years, before being abandoned after springing a leak.

So can the air be harvested for moisture on any useful scale? The atmosphere contains around 10.5 billion acre-feet of water at any one time, six times more than all the world's rivers. Of this, about 2 percent has condensed into water droplets in clouds and will mostly fall as rain before too long. The remaining 98 percent is still in the form of water vapor. It is a tantalizing prize, and some outlandish machines have been proposed to capture it.

In the 1930s a French meteorologist called Bernard Dubos proposed building a chimney 2000 feet high with a fountain at its base to create an updraft of humid air that would, he believed, saturate the air above and generate rain. The machine was never built. But military people have from time to time experimented with manipulating the electrical charges in clouds to encourage the formation of raindrops. In the mid-1990s, I came across one such, a small Russian unit doing just this in the badlands of Jordan, on the way to the Iraq border. Few results have ever been published. In the 1940s, South Africa's chief meteorologist, Theodore Schumann, proposed erecting an electric fence 150 feet high and 2 miles long on the top of Table Mountain outside Cape Town to ionize the air and condense millions of gallons of water a day from the atmosphere. Again, it never happened.

Some have even suggested using sound to capture moisture from the air. On a cool, still night, the air can be so saturated with moisture that even modest air movements, such as sound waves, can condense the moisture and produce raindrops. In the mountains of Yunan in southern China, villagers have a tradition of yelling loudly in the hope that it will stimulate rain. The louder they shout, it is said, the more it rains. This gives interesting scientific credence to the African notion of the rain dance, once seen as the epitome of superstition.

An interesting new idea that demonstrably works is to condense moisture from hot desert air in a greenhouse, using seawater as the cooler. It works best where there is a handy cold offshore current that can be pumped to the greenhouse. Charlie Paton, a British engineer, has produced working "seawater greenhouses" on Tenerife in the Canary Islands and in Abu Dhabi and Oman. The greenhouses require complex systems for heat exchanging to have the full effect, but they are in essence giant dew-making machines. The Royal Institute of British Architects called the design "a truly original idea which has the potential to impact on the lives of millions of people living in coastal water-starved areas round the world."

Paton's Abu Dhabi greenhouse has been growing cucumbers, tomatoes, and flowers in the desert since 2002. It delivers 10 gallons of water a day for every 10 square feet of greenhouse—that is more water than falls in the rain of an average rainforest. As well as producing water, the heat exchangers air-condition the greenhouse, so it makes cool air in the desert, too. As a result, the crops need much less water. In the Abu Dhabi greenhouse, this means ten times as many crops can be grown with the same amount of water.

⌒

And then there is rainmaking. To many, seeding the clouds to make rain seems outlandish. But in fact scientists have been active in this business for more than half a century. The idea is to encourage the formation of raindrops by spraying billions of tiny particles into the clouds. These, so the theory goes, will form the necessary nuclei around which water droplets will form and coalesce to make raindrops. Nature does this mostly with tiny salt crystals from sea spray. Scientists mimic the process, usually with silver iodide crystals. Successes have been claimed, but proof is hard to come by. That is understandable. Who can predict what would have happened without the seeding?

But practitioners are convinced, and surprisingly large sums of money are spent on operational cloud-seeding. It is practiced currently in twenty-four countries and ten American states, from North Dakota to Texas. In North Dakota alone, planes spent more than 600 hours in the air spraying clouds in 2003. The Idaho state power utility invested almost a million dollars seeding clouds during the winter of 2004–2005. It claimed the money was well spent. A previous operation had increased snowfall by 16 percent, increasing spring runoff into the state's rivers. This in turn raised hydroelectric power generation and reduced the utility's need to buy power from neighboring states.

In Australia, New South Wales plans to seed clouds to produce snowfall and maintain ski runs in the Snowy Mountains. But Israel claims the most successful and long-standing cloud-seeding operation in the world. According to the country's rainmaking guru, Daniel Rosenfeld, it works especially well in winter clouds rolling across the north of the country, over the hills of Galilee. He claims that spraying typically raises rainfall from clouds by 15 percent and produces 40,000 acre-feet of water a year—adding 3 percent to the country's overall water budget.

Cloud-seeding is good politics in a drought. As the planes take off and spray the sky, authorities can at least say that they are doing something to fill faucets and irrigation canals. The Indian state of Karnataka is a cloud-seeder. So are the African governments of Morocco and Burkina Faso. All claim success. But the scientific basis for the claims is not clear.

While cloud-seeders suffer from an inability to prove success, they also suffer from public fears that they could be all too effective. For one thing, cloud-seeding can be an act of war. The United States secretly seeded clouds over Laos and North Vietnam in the 1960s to waterlog the trails of human donkeys taking supplies for guerrillas south during the monsoon. And sometimes seeding has been accused of causing unwanted floods. Most notoriously, Operation Cumulus, a secret military rainmaking experiment over southern England, may have triggered huge storms on Exmoor in the summer of 1952. A rush of water and mud hit the Devon village of Lynmouth a few hours after one spraying, killing thirty-five people. The truth of cause and effect can never be known, but according to documents unearthed years later in the Public Record Office, Operation Cumulus was abruptly suspended shortly after the disaster. Perhaps officials knew enough.

There are other problems. Cloud-seeding, if badly done, can stop the rain.

Too many crystals thrown into clouds will cause the moisture to form too many droplets, none of which become big enough to fall as rain. And even if seeding generates local rain, the question remains whether that will mean that less rain falls somewhere else. Jordan, one of the driest and most water-stressed countries in the world, fears that Israeli cloud-seeding may be emptying rain clouds blowing east from the Mediterranean before they cross the Jordan Valley. Canada's prairie provinces have similar fears about U.S. operations in North Dakota.

China spent a quarter-billion dollars on cloud-seeding between 1995 and 2003 and, so its meteorologists claim, produced 160 million acre-feet of extra rain—half the flow down the Yellow River—in that time. As well as making rain, clouds were seeded to prevent hail formation, damp down forest fires, disperse fog, and even bring down summer temperatures. In 2003 alone, 30,000 staff members, 3800 rockets, 7000 shells, and 30 planes were employed in the task. But across the plains of eastern China, there have been dark warnings of rain theft. In the middle of another dry summer in 2004, Pingdingshan city sent up planes to spray passing clouds. Some 4 inches of rain fell on the city. But 90 miles downwind, Zhoukou city received only a fifth as much. The city fathers concluded their neighbor had emptied the clouds of the rain that was rightfully theirs. Maybe so, but I was a few hundred miles away at the time, and it was raining where I was too, without the benefit of seeding. So who knows?

In one of the world's driest deserts, they used to harvest fog. The harvest site looked like a volleyball court for giants: seventy-five large sheets of plastic mesh suspended along a remote hilltop in the Atacama Desert of northern Chile. It does not rain here for years on end, but fogs roll in regularly from the cold offshore current of the Pacific Ocean, and the plastic sheets were getting wet by capturing the moisture from those fogs. Like giant sails pointing into the wind on the ridge, the polypropylene nets collected two thirds of the moisture in the fogs.

Each sheet, measuring 40 feet by 10 feet, took 40 gallons of water a day. Tiny droplets accumulated on the mesh until they formed single large drops that ran down into a trough that flowed to Chungungo, a small shellfishing town of 350 people that previously depended for its water entirely on tankers driven from 50 miles away. The project provided an average of 4000 gallons of water a day.

After a seven-year research project, the nets on El Tofo ridge were hailed as the first new method of providing drinking water in over a century. "Until we set up the water system, the town was dying," said Bob Schemenauer, a Canadian who helped Chilean researchers mastermind the technology in the 1990s and has since spun off his own organization, FogQuest, to develop it. "But with a secure water supply, people began to return, and new houses were built."

Well, that was what he said when we spoke in the 1990s. Five years later he was wiser. The town had indeed grown, so much so that the sheets could no longer meet its needs. There was plenty of room on the ridge, but instead of putting up more sheets, the government decided to install a pipeline to the resurgent town, at a cost of $1 million. After that, neither the local authority nor the townspeople maintained the sheets. The system is now defunct, the plastic sheets torn and abandoned.

But the idea is catching on elsewhere. All along the Pacific coast of South America, communities have put up fog-catching screens, usually to provide water for new tree plantations. Often, once established, the trees collect the fog for themselves on their foliage, recreating fog-based ecosystems in the desert like the ones that flourished before settlers chopped down the forests.

Schemenauer has moved on. He has developed his fog-harvesting system for other arid lands, from the Caribbean island of Haiti to Namibia, Nepal to Yemen, and Guatemala to Eritrea. In Oman, where overpumping has wrecked groundwaters and desalination plants are the main source of water, there are fogs on around eighty days every year. When Schemenauer put up nets there, he got ten times the yield that he managed in Chile. In Yemen, where the central plateau is set to run out of underground water by 2010, he wants to erect nets that could replenish ancient cisterns carved into the hillsides. Israeli scientists are experimenting in the Negev Desert and on the Golan Heights. Other tests have been going on in the Peruvian desert and on Tenerife in the Canary Islands. Undoubtedly the technology works. But if the experience in Chile is any guide, the problem is finding a way of encouraging locals to use it when governments, at the end of the day, will never let them die of thirst.

Nature is a good fog-catcher too, and may have some tips for would-be fog harvesters. On El Hierro, another Canary Island, people harvested fog droplets from the leaves of trees until a hundred years ago. Perhaps this practice was the origin of a report by Pliny the Elder, in Roman times, of a Holy Fountain Tree growing on the island. In Namibia, meanwhile, British zoologists recently discovered a beetle in the desert that has evolved a bobbled upper surface to its body with a pattern that is supremely efficient at capturing moisture from passing fogs. The hexagonal pattern of tiny peaks and troughs appears to push tiny droplets together to form larger droplets, which then roll off the beetle's back and into its mouth. The scientists, headed by Andrew Parker, of the University of Oxford, rigged up a prototype fog-catching surface based on the beetle's design, which captured five times more water than Schemenauer's netting. So they patented it, and several companies are vying to make fog-collecting devices that can be put on the roofs of buildings, or even tents.

Is desalination of seawater the answer to the world's water woes? Some say so. Distilling seawater by boiling it and collecting the water vapor is an age-old activity. But modern distillation technology was developed by the U.S. Navy for operations on remote Pacific islands during the Second World War. Following that, large-scale distillation for public supply took off in the water-poor Gulf states, where they have plenty of oil to provide the necessary energy.

Today global desalination capacity is about 8 million acre-feet a day—roughly 3 percent of the global tapwater supply, though only a tenth of 1 percent of total water use. Most of the global capacity is still in the Gulf states. Saudi Arabia alone accounts for one tenth of world output, and in 2004 announced plans for six more plants costing a total of $5 billion. Islands where summer tourists have overwhelmed local water supplies have until recently made up most of the rest. In Mediterranean Europe, Malta gets two thirds of its drinking water from desalination. Greek islands like Mykonos have been doing the same for years, as have Caribbean islands such as Bermuda, the Caymans, Antigua, and the Virgin Islands.

Four fifths of the world's total desalination capacity still uses distillation, but since the 1970s, an alternative technology has grown in popularity. Reverse osmosis forces water repeatedly through a membrane that filters out the larger salt molecules and lets clean water through. Both technologies require large amounts of energy, whether to evaporate the salty water or to force it through the filters. Until recently, it cost several dollars to produce 265 gallons of unsalty water—typically a hundred times more than conventional water supplies. But costs for reverse osmosis have come down, and more cities are buying into the technology.

Tampa Bay, Florida, and Santa Cruz, California, have both taken the plunge, and more reverse-osmosis plants are slated for Houston, Texas; Cape Town, South Africa; and Perth, Australia. In Spain a new government elected in 2004 swiftly abandoned its predecessor's plans to pump water cross-country from the wet north to relieve the arid south, in favor of twenty reverse-osmosis plants. They are expected to meet slightly over 1 percent of Spain's total water needs.

The cheapest desalinated seawater is now in Israel, where one of the world's largest reverse-osmosis plants has been built on the Mediterranean coast at Ashkelon. Israeli water economics are notoriously opaque, but the government claims to be able to deliver water at around fifty U.S. cents per 265 gallons, around a third of the production cost in Saudi Arabia. (More pertinently for Israelis, it compares with the thirty cents it costs to pump freshwater from the Sea of Galilee to Tel Aviv and the two dollars to buy water by the tankerload from Turkey.)

Such prices are encouraging cities in less extreme circumstances, and in cooler and wetter climates, to join the reverse-osmosis revolution. During 2004, China announced plans for a giant plant for Tianjin, the country's third largest city, where water shortages have been endemic for years. Even more surprising, Britain's Thames Water announced that it will build a $400 million reverse-osmosis plant on the Thames estuary in east London to process water during droughts. It will be able to produce up to 120 acre-feet a day, enough to meet the domestic needs of almost a million people.

The boom in desalination is beginning to alarm environmentalists. One problem is what to do with the salt extracted from the seawater during the process. It emerges as a vast stream of concentrated brine. Most plants, naturally enough, dump it back into the sea. But this salty wastewater also contains the products of corrosion during the desalination process, as well as chemicals added to reduce both the corrosion and the buildup of scale in the plants.

Maybe this pollution can be fixed technically one day. But what can't be fixed is the huge energy demand of desalination. A typical reverse-osmosis plant consumes six kilowatt-hours of electricity for every 265 gallons of water it produces. Most of the power, inevitably, comes from burning coal, oil, and other fossil fuels. So while desalination could conceivably become a viable source of drinking water in coastal regions around the world in the coming decades, it would be at the expense of an extra push toward climate change.

It is also hard to see desalination ever penetrating the agricultural market, where the majority of the world's water is currently used. At the end of the day, desalination seems like an expensive high-tech solution to a global water problem that is overwhelmingly caused by wasteful use. Like enormous engineering projects for shifting water around the planet, it is a supply-side solution to a demand-side problem.

IX

When
the rivers
run dry...

we try to catch the rain

28

Catch the Rain

Lian Jianmin is a smiling, worldly, chain-smoking raconteur. He was telling me his life story. "I was a seventeen-year-old just out of high school in Tianjin in the east. During the Cultural Revolution in the late 1960s, when students were sent to the countryside to learn about peasant life, I went to work in a village in the hills of Gansu, in western China. I saw they all had these cellars under their houses, where they collected the rain from their roofs." Many of the cellars were very old and had silted up, he said. People preferred water from wells. But then there was a drought, and the wells went dry. Villagers without cellars had to walk 5 miles or more to find water. "Even the birds followed the water tankers," he said. And afterward, everyone started digging out the cellars.

The sandy hills of Gansu are one of China's poorest and most desolate regions. They have no rivers other than the Yellow River down in the valley, and the water in the wells is often salty these days, and prone to run dry. But they do have a little rain and a long tradition of catching it before it can run away. Some of their water cellars are said to be eight hundred years old. The tradition may go back two thousand years. Lian said he had met a farmer who was eighty years old: "When he rebuilt his house, which he'd lived in all his life, he discovered a cellar underneath that still contained water. And the quality of the water was very good."

Lian was so captivated by the structures for harvesting rain that he decided to make their development his life's work. The city boy from the east has never left Gansu. Today he sings the praises of the local peasants and their cellars to anyone who will listen. These days, that includes a large number of foreigners who have come to learn the wisdom of these hills. The world has come to him. Lian, now an engineer with the province's water resources bureau, entertains a growing galaxy of foreign water experts. "We had people from thirty countries here last year," he told me.

This large, arid province, straddling the middle reaches of the Yellow River, is solving its rural water problems without waiting for any promised benefits from distant water delivered by the south-to-north river diversion—a scheme Lian sees as wasteful and unnecessary. And he and his fellow researchers practice what they preach. At the head office of the agricultural development research center in Dingxi, where we talked, they have almost nine hundred underground cisterns that provide all the water for irrigating their greenhouses and research plots.

An estimated 2 million people in Gansu get most of their water for drinking and growing crops by harvesting the rain—a figure that is rising all the time. These days the cellars catching water for domestic use have been augmented by an increasing number of cisterns in the fields for growing crops. Big enough for a man to stand in, they usually have a distinctive bell shape and typically hold 16,000 gallons of water. They are usually lined with cement, though some say the old cellars lined with mud bricks give better-tasting water.

Every household has at least one cistern. Traveling to the villages around Dingxi, I met a sheep farmer named Zhujia Zhuang. He had piped water but kept a backyard cistern to water a small vineyard and a second one hooked up to a solar heater in a shower cubicle. An Yueying, also a farmer, showed me around her tiny but immaculate farmhouse. She relied on the rain for all her water. A cistern outside her house caught water from the gently sloping yard, and another collected water running off a hillside and down a small concrete channel. "I built them myself about twenty years ago," she said. She was growing alfalfa for her cattle in a field, and pear trees and vegetables in the yard. It was the end of the dry season, but there was still water in both cisterns.

Was the water safe to drink? Usually it was all right, she said, but to be sure, she boiled it first. As we spoke, she filled a battered old black kettle from the cistern in the yard and hung it up above a giant shiny dish in the yard pointed toward the sun. I had at first taken this for a big TV satellite dish, but in fact it was concentrating the sun's rays to heat the kettle. A while later, the kettle came to the boil.

Later, over lunch, I met Zhang Zhenke, the chubby director of the Dingxi county water conservancy bureau. Most of rural Gansu, especially in the east of the province, drank from cellars, he said. "We have 200,000 water cellars in this county alone," he explained. "A quarter of them are privately built, and the rest were initiated by the government." For farming, many people set aside small fields as a catchment area, laying plastic sheeting on the ground to insure they lose no water.

In many parts of the world, local officials would be embarrassed to have only a tenth of their population connected to piped water, but Zhang was proud of the achievement. "We already have more rainwater harvesting here than anywhere else in China, probably the world," he said. Cisterns, he declared, were the key to the future prosperity of the region. The policy now was for every household to have a concrete catchment of 1000 square feet to fill two cisterns. "Eventually we want four for every household," he said. "There would be one for drinking water for a normal year, one for a drought year, and two for irrigating crops. We are very proud of what we have done here to survive. We have some of the worst conditions in the country. Without the cisterns, people could not live here."

In China, it was Chairman Mao and his Cultural Revolution that began the revival of the ancient tradition of rainwater harvesting. In India, it has been a mixture of swamis and scientists, schoolteachers and even policemen. Haradevsinh Hadeja is a retired Indian police officer. He loves playing cricket for his village team. When we met, he had a broken arm caused by a fast bowler from the next village. But win or lose at cricket, his village of Rajsamadhiya, in the backwoods of Gujarat, always excels at water.

His fellow villagers say Hadeja is a near-magical diviner of water, and you can see why. Around the village, he has transformed a desertlike landscape of desiccated fields and empty wells into a verdant scene of trees, ponds, full wells, and abundant crops. With no piped water, most of the other villages in the area rely on tankers to provide drinking water for much of the year. They have little left to irrigate their crops. But here, says Hadeja, "We haven't had a water tanker come to the village for more than ten years. We don't need them."

He worked this miracle by catching the monsoon rains. Not, like the Chinese, in purpose-built cellars and cisterns, but in ponds. And the villagers don't use the water directly from the ponds. They allow it to percolate into the soil to refill underground water reserves and replenish their wells. "There is no more rain than before. We just use it better. We don't let it wash away," Hadeja says. The village has twice as much water as before, and wells find water at only 20 feet down, whereas once the water had to be hauled up from more than 100 feet. The contrast between Rajsamadhiya and surrounding areas where water tables are falling is extraordinary.

I went to look. The heart of the village appeared conventional enough: a gaggle of single-story houses leading from a small square out toward the fields. But on the paths there were thousands of fruit trees, where most villages are treeless. Under their shade were piles of mangoes and watermelons. And out among the small fields growing wheat and vegetables and groundnuts, there were the ponds—lots of them. "We have forty-five water-collecting structures altogether," said Hadeja as we walked past a line of women washing clothes in a pond. "This one gets its water from land up to 3 miles away," he said.

The ponds are arranged along the routes that the monsoon water takes as it drains through the village. Rather than trying to get rid of the floodwater, Hadeja has redesigned the village's drainage to slow the water's passage long enough for it to collect in specially dug ponds. The water passes from one pond to the next in a slow cascade, seeping through the soil to refill the aquifer all the way.

Hadeja's second innovation has been to manage how people use the water. As a former policeman he has some authority, and though all decisions are taken by the village council, his word has obvious force here. (In fact, he told me quietly, the police and other authorities never come to the village now. They leave it all up to him.) Under Hadeja's law, nobody is allowed to take water directly from the ponds, and farmers are banned from growing the thirstiest crops, like sugarcane. "There is no point in catching more water if we only waste it," he said.

News about this remarkable village has spread around India and beyond. One foreign scientist brought satellite images of the village that showed hidden cracks in the geology through which water was flowing. Hadeja slowed the flow by plugging the cracks with concrete. But mostly, as in Gansu, water scientists are coming here to learn about rainwater harvesting, though the water diviner insists that there is little to learn. "I am an uneducated person," he said. "I saw that people were leaving the village and I wanted them to stay. That meant finding more water. So I tried to catch the rain."

And in truth, this is no miracle. Hadeja has tapped into an old tradition and developed it. In India, you can still see abandoned ponds and lakes dotted across the countryside and on wasteland in cities. Until the early nineteenth century, much of India was irrigated from shallow mud-walled reservoirs in valley bottoms, which captured the monsoon rains each summer. The Indians called them *tanka,* a word the English adopted into their own language as "tanks."

Most of the tanks were quite small, covering a couple of acres at most, and irrigating perhaps fifty acres. Farmers scooped the water from the tanks, diverted it down channels onto fields, or left it to sink into the soil and refill their wells. The tanks served other functions, too. Some were stocked with fish. All were prized for the silt brought into them by the rainwater. Farmers guarded the slimy, nutrient-rich mud in their tanks almost as much as the water. They dug it out to put onto their land and turned silted-up former tanks into new farmland.

Overall, across India, researchers estimate there are around 140,000 tanks, either still in use or abandoned. Tamil Nadu has the most, approaching 40,000, covering several percent of the land surface of the state and still irrigating around a 2.5 million acres. Karnataka estimates it has 35,000. Every region has its own design. In the Thar Desert of northwestern India, people channel runoff into manmade desert depressions called *khadin,* creating wet soil for planting wheat or chickpeas.

The system thrived until the British took charge in India. Though full of admiration for some of the grand Indian water structures on rivers, British water engineers largely ignored the village tanks, apparently not realizing that they were the way India fed itself. Tanks passed into a kind of forgotten underworld, used while they served a purpose but unrecognized by officialdom and rarely repaired or cleaned out. As the British and later the Indian government promoted more modern water-gathering technologies, they gradually fell into disuse. But today, as the formal irrigation systems established on the Western model fail across the country, and as farmers are having to pump from ever greater depths to retrieve underground water, the old tanks are starting to be restored.

All across India, groups are harvesting the rain either for direct use or to revive underground water reserves. Besides tanks and *khadin,* there are also check dams. These are barriers constructed in small streams and gullies to hold up the monsoon rain long enough for it to percolate underground. In Rajasthan, a government scientist, Rajendra Singh, gave up his job, taught himself traditional water-harvesting skills, and went out to the edge of the Thar Desert, where villages were dying for want of water. He encouraged the locals to install check dams. Now his movement has 4500 of them collecting water in several hundred villages, forty-five permanent staff members, and a grant from the Ford Foundation.

Whole landscapes are being transformed. In Limbadia, a village in western Gujarat, the water table was 500 feet down and falling fast until villagers built a series of check dams. Soon afterward several wells began spurting water at the surface. Tushaar Shah, at the International Water Management Institute in Gujarat, estimates that across Rajasthan, some 2500 square miles of land are being newly managed to capture the rains, "with dramatic impact on groundwater recharge and the revival of dried-up springs and rivulets." Water tables have risen so much in Rajasthan that five ancient desert rivers—the Ruparel, Arvari, Sarsa, Bhagani, and Jahajwali—have returned to the map.

In Karnataka, on the plains west of Bangalore, I visited a group called the BAIF Development Research Foundation, which has helped farmers dig 350 ponds across four valleys near the town Adihalli. Water flows from one pond to the next in a slow cascade. The result is more water in village wells, year-round farming of grains, and improved yields of cash crops like coconuts and chiles, cashews and mulberries, vegetables and rice. Incomes have doubled and sometimes quadrupled.

"There used to be water on the land ten days a year; now they have it all year round," said the director of the project, a Gandhi disciple named G. N. S. Reddy. One local farmer told me with delight, "I can irrigate my vegetables from the pond once a week, and afterward there is still water in the pond." As Reddy and I walked back to the road, with birds singing in the trees, we met women goatherds who had walked several miles to the ponds. "Where we live, there is no water. So we come here. They let us use the water," one told me.

I met a husband and wife who a few years before had given up farming and become migrant laborers for much of the year—pretty much the bottom of the heap, even in India—after their 325-foot-deep borehole went dry. But now, thanks to the local network of ponds, the water table was back to 130 feet and they were home again. When we dropped by, they were wielding a long hose-pipe to irrigate rice and gherkins. Ratnama, the wife, said with a broad smile, "We used to steal fodder from other people's land for our animals. Now other people steal ours. We've built a new house and sunk another borehole. We are not begging anymore. We can even get loans."

Rainwater harvesting is becoming a widespread social movement, uniting many strands of Indian society. Pandurang Shastri Athavale, a Vedic scholar in western India known to his followers until his death in 2003 as Dada, preached a simple life, in which commonly owned resources such as water were revered and their protection was seen as an act of devotion. With this creed, Dada persuaded tens of thousands of villagers to construct low mud walls on their fields to divert the monsoon rains directly down their wells. Some 300,000 wells have so far been adapted to receive rainwater.

"Even cities can do it," says Sunita Narain, the director of the Delhi-based Center for Science and the Environment, an outspoken advocate. "In parts of Delhi where old tanks and ponds have been cleared of garbage and refilled with water, the water tables are rising." Delhi could, if it got organized, obtain a third of its water from harvesting the rains, she says. In Bangalore, in India's Silicon Valley, they are trying to boost the aquifers by rehabilitating the city's sixty ancient lakes. The city's water is among the most expensive in Asia, be-cause to keep the city supplied, the authorities pump 400,000 acre-feet up 1600 feet from the distant and drying Cauvery River. The Bangalore architect S. Vishwanath says, "The city gets about 800,000 acre-feet of rainwater a year. We need to use that instead."

Vishwanath is part of a movement to revive the city's own water reserves called the Rainwater Club. I saw houses, offices, apartment complexes, schools, even a planetarium that were catching water from roofs, gardens, paved areas, and parks, thanks to its work. Vishwanath's own eco-house captures enough water from the roof during the monsoon season to flush toilets and run the washing machine for most of the year. An evening's rain can be enough to run the house for a week. From the roof beside the cistern, Vishwanath pointed down the street to a city water tower. "They supply water about every four days. I have water from my roof anytime I want it," he said.

But as heartening as these individual initiatives are, the roots of the Indian rainwater-harvesting revolution lie in the communal rehabilitation of aquifers. Shah set up his groundwater research center in Gujarat partly to learn the lessons of people like Hadeja, Singh, and Dada. He says the vital factor in their success is that the initiatives happen at the village scale. Few individual farmers can successfully catch their own rain and store it underground; it would quickly dissipate into the wider aquifer. But when an entire village does it, the effects are often spectacular. Water tables rise, dried-up streams flow again, and with more water for irrigation, the productivity of fields is transformed.

Shah says the rainwater-harvesting movement is "mobilizing social energy on a scale and with an intensity that may be one of the most effective responses to an environmental challenge anywhere in the world." It is, he points out, completely autonomous from government. "It emerged on its own, found its own source of energy and dynamism, and devised its own expansion plans."

By some estimates, 20,000 villages in India are harvesting their rains. Not much, perhaps, in a country that boasts a million villages, but it is a start. And however the "social energy" is created—whether by the force of a personality like Hadeja's or through religious devotion or a Gandhian ethos—some kind of communal water ethic seems to be the magic ingredient. That and a belief, whether expressed in religious or secular terms, that, as Dada put it, "If you quench Mother Earth's thirst, she will quench yours."

29

On the Grapevine

There has never been a blueprint for catching the rain. You can catch it on rooftops or on agricultural terraces, by diverting flash floods into ponds, by blocking gullies, or by putting low earth embankments across hillsides or around individual plants. But harvesting the rain was once a worldwide technology on which hundreds of millions of people depended. Every locality had its own systems. Almost everyone did it. English country houses caught the rain from their roofs to water the flowerbeds. So did palaces from India to the Middle East. African villagers harvested water from the leaves of trees in their kitchen gardens. American Indians erected low stone walls and weirs made of brushwood across the Sonoran Desert to divert stormwaters onto their fields. For small tribes and communities in particular, rainwater harvesting made much more sense than the larger river-diversion structures of "hydraulic civilizations."

Many countries, particularly in the Middle East, have almost apologetically continued harvesting the rain in defiance of Western engineering norms. Jordan constructs earth dams in the desert to increase soil moisture for groves of olives, pistachios, and almonds. Yemenis have captured floodwaters in wadis for centuries and exported the idea in the early twentieth century to the East African states of Eritrea, Ethiopia, and Somalia, where this is now the dominant form of irrigation. Tunisia still grows 100,000 olive trees with *meskats,* small rain catchments as big as tennis courts and surrounded by earth embankments, or bunds, that channel water to a cropping area of about half the size. They tap rainwater and springs in the wet north and route it around precipitous hillsides and through tunnels toward the sunny, arid south, where the precious liquid grows bananas and grapes. A similar system continues to water orchards in parts of Afghanistan.

In the desert sands of Egypt, the ancient Bedouin technique of cutting cisterns into the rocks to catch and store water for drinking and irrigating vegetables has undergone a dramatic revival. Thousands of new cisterns have been cut.

The inhabitants of small islands surrounded by salty seas know better than most the perils of a finite supply of fresh water. Madeira, a holiday destination in the warm waters of the Atlantic Ocean west of Morocco, has few rivers. But it does have a network of 1500 miles of beautifully constructed stone water channels that cross the entire island. Known as *levadas,* the channels were constructed by African slaves for Portuguese sugar plantations. The British territory of Gibraltar, on the southern tip of Spain, which is politically if not quite geographically an island, has one of the most intensive rain-collection systems in the world, channeling water from collecting areas on the giant rock that dominates the enclave into tanks excavated into the rock. Across the Atlantic in Bermuda, new houses are required by law to include equipment for collecting rainwater from the roofs.

———

If any moment was pivotal to the rediscovery of rainwater harvesting in the West, it was when an Israeli archaeologist, the late Michael Evenari, from the Hebrew University of Jerusalem, stumbled onto hillside water channels in the Negev Desert. Evenari had been excavating settlements of the Nabateans, who controlled the caravan routes from Arabia to Rome via Mediterranean ports in Gaza around two thousand years ago, trafficking in gold, ivory, spices, frankincense, and myrrh. This was big business, the ancient equivalent of running an oil pipeline. The Nabatean kings were probably the three kings of the nativity story, and they ran six imposing desert cities, including the famous "pink city" of Petra in modern Jordan. But how, Evenari wondered, were they fed and watered?

As he excavated, Evenari realized that these traders had caught the rain as it fell in occasional storms on the hills. Hydrologists estimate that a single acre of Negev Desert can capture as much as 107,000 gallons of water over a year. Cleverly managed, this was enough to moisten the dry soil and grow olives and wheat and vines. It seemed to be a model of how to live in the desert. Evenari was so struck by the idea that he recreated an ancient Nabatean farm below the ramparts of Avdat, one of the ruined cities in the Negev.

When I paid a visit to the Avdat farm, it hadn't rained for six weeks. All around was barren wasteland. But on the farm the soil was damp, a field of wheat was growing, and almond and pistachio trees were in leaf. "We have tried to grow the crops mentioned in the Bible," said Pedro Berliner, a researcher. "And most of them will grow here in the desert if we harvest the rain." It was a stunning sight. And as I have spoken to practitioners of rainwater harvesting around the world since, I have discovered how important the Negev experience has been in their work.

Vedic scholars from India, it turns out, first learned of rainwater harvesting from the Nabatean work. Other Indian advocates made similar pilgrimages to Israel back in the 1980s. It was around then, too, that Bill Hereford, a worker for the aid agency Oxfam, came across Evenari's desert farm while on a sabbatical from Burkina Faso, in West Africa. On his return, he encouraged African villagers to try to revive abandoned desert fields on the edge of the Sahara by laying low stone walls on the hillsides. It worked. The walls captured water long enough for it to soak in and nourish crops.

The news spread from village to village and eventually into neighboring countries and across much of the West African bush. Thousands of acres across the arid Sahel, which once seemed destined to turn to desert, now have crops and trees growing where before nothing grew. Where the rain has been harvested, there has been a 70 percent increase in yields of local grains such as sorghum and millet, as well as more trees for firewood and more grassland for livestock. Chris Reij, of the Free University, Amsterdam, says rainwater harvesting is the leading cause of a widespread greening of the Saharan margins in the past fifteen years.

There have been other notable instances of the cross-fertilization of the rainwater-harvesting idea. In the Machakos district of Kenya, the Akamba tribe was in a bad way sixty years ago. A British administrator called the hills where they lived an "appalling example" of environmental degradation: "The inhabitants are rapidly drifting to a state of hopeless and miserable poverty and their land to a parched desert of rocks, stones, and sand." But rather than accept their fate, the Akamba people sought out new ideas. Elders remembered seeing tanks and hillside terraces to catch water in India, where they had served in the British colonial forces during the Second World War. They decided to try it back home.

They have since dug ponds and terraced hillsides to such effect that today, despite a huge surge in the Akamba population, their farms are producing more, their people are wealthier, and their landscape is greener. Researchers call this transformation the Machakos miracle. Desertification has been put into reverse through catching the rain.

Meanwhile, Oxfam has encouraged rainwater harvesting on the West Bank, where some 50,000 Palestinians catch rain on their roofs during at least part of the year. Such systems have a long tradition here. The Jordanians used to insist on it in towns like Ramallah when they were in charge, before 1967. But now it is a means to get around Israeli restrictions on drilling new wells. Jad Isaac, the director of the Applied Research Institute in Bethlehem, estimates that Palestinians take 8000 acre-feet of water a year from rooftops in Gaza and the West Bank for both drinking and irrigation.

⸺

What is the potential? "Millions of acres of the dry parts of the world were once used for water harvesting. In the twentieth century there has been a steady decline, but the twenty-first century is likely to see a huge revival," believes Dieter Prinz, of the University of Karlsruhe in Germany. Social and economic conditions may have changed, "but the advantages of water harvesting remain valid, and farmers in dry areas will have to use them if they want to be able to master their future."

Steve Halls, the director of the UN Environment Program's International Environment Technology Center in Osaka, agrees. Two billion people across Asia could have their water shortages relieved by rainwater harvesting, he says. "You can take basic rainwater-harvesting systems, which look small, and scale them up till they are helping millions of people. It is a technology that can be used right around the world, in urban as well as rural settings. As long as you have rain, all you need to do is plan where to build the holding tanks."

But many believe that the real potential for rainwater harvesting to transform lives lies in rural Africa. This is where unmet water demand is greatest, where huge areas have no clean running water, and where millions of women walk for miles every morning to collect water from fetid pools and rivers or unreliable wells. It is where people go hungry for want of water to irrigate their crops. It is where the residents of squatter colonies in cities pay two dollars or more to receive water from tankers run by private vendors. Most of the increase in world demand for water in the coming half-century will be in Africa.

⁓

Johan Rockstrom, a Swedish hydrologist, calculates that conquering existing hunger and coping with an estimated 3 billion extra people will increase global annual water consumption by 4.5 billion acre-feet, or 80 percent, by 2050. Where will that water come from? Extra dams cannot get anywhere near meeting that need, he says. There isn't the money. And even if there were, most rivers in farming regions are already running dry, and the good dam sites are mostly taken. At top weight, they could supply only an extra 650 million acre-feet of water per year.

Meanwhile, most of the world's underground water reserves are ruled out because they are already overpumped. Their use should be scaled back rather than increased. Novel water sources like desalination may help in some places with drinking and industrial water—but not irrigation. So the future of the world's food depends on catching the rain. If Africa is to feed itself in 2050, it will have to increase rain-fed farming fivefold, Rockstrom says. "We need individual farmers to catch the rain," says David Molden, of the International Water Management Institute. There is no alternative.

Hearing these words, I thought of Jane Ngei, a poor farmer with a young family I met on a back road in Machakos in Kenya. She had never heard of the Machakos miracle, still less of her ancestors' visits to Indian terraces or a global movement in rainwater harvesting. Yet she showed me proudly how, with a wheelbarrow and a shovel, she had dug herself a pond to catch water that poured down the road by her farm after rains. She used the water to irrigate a kitchen garden of corn, vegetables, and fruit trees. The extra money she got from selling her produce in the local market was paying for her children to attend school. It was simple. It had transformed her life.

Ultimately, the big numbers of the hydrologists come down to this. Africa needs millions of Janes if it is to remain fed and watered.

30

Unfailing Springs

Shallalah Saghirah is a small, shabby village in northern Syria. Its name means "little waterfall." But when Musa Oqlah came here from the south in 1928, the land was deserted and dry, a frontier of dry Bedouin goat pastures. When he planted barley, the rains were not good enough and the crop died. He was on the verge of moving on when he noticed that in one spot the barley grew around a seep of water. He started digging and found what seemed to be a well. He stayed and drew water from the well to irrigate his barley. The year is etched on the memory of everyone in the village today as the founding date of their community.

A decade later, in 1938, Oqlah's well ran dry. Again he thought of moving on. But first he did some more digging in the gulley behind the village. What emerged staggered him. He found more wells, and discovered that the wells were part of a chain, connected at their base by a tunnel 1300 feet long, which tapped underground springs up in the hills. Oqlah had discovered what is known in these parts as a *qanat*. He and his sons cleaned out the wells and the tunnel and constructed a stone surface channel and reservoir to take the water to the middle of the village. The *qanat* has flowed ever since.

Shallalah Saghirah has about twenty households today. All the seven elders of the village are cousins, grandsons of Oqlah. They have no electricity, and their only source of water is the underground tunnel. Nobody knows how old the tunnel is. Ismail Hasoun, one of the cousins, says, "It is Roman; it is four thousand years old." But he doesn't know. At least 1500 years seems a likely guess, say local researchers who took me there. It is, however, an unfailing spring. After a renovation in 2000, the village doubled the discharge of the system. Today it pours into the small reservoir at a rate of about a quart a second, as regular as clockwork, whatever the season.

Hasoun is an elderly man now, his weathered face peering from beneath a kaffiyeh. But he clambered like a mountain goat up the gulley behind the village to show me the chain of seven wells along the tunnel's route. He pointed out the unlikely rocky outcrop from which the water first enters the tunnel from underground springs. He had explored the tunnel for the first time during the recent renovation, and said that near the head it branches into several sections, apparently to tap more of the aquifer. He mimed how he had descended the wells and cleaned the tunnel. It was clearly hard work, but essential for the maintenance of this mysterious artery that sustains his village.

Hasoun's grandfather divided up the water from the tunnel between his five sons. Each got a day of the water flow in turn. The fields are still owned by their descendants, and the share-out continues much the same. Today they grow figs, mulberries, vegetables, and barley. As we talked, another cousin was plowing the fertile soil with the village tractor. "The other villages in the valley have to buy water from tankers or the government wells," Hasoun said. "And their water isn't as good as the water we get from the tunnel."

Researchers have mapped around 250 *qanats* across Syria. They remain the main source of water in a few dozen villages. But most have been abandoned, either because they have filled with silt or because pumped boreholes have lowered the water table and dried them up. One tunnel, supplying Turkish baths in the ancient Byzantine city of Aleppo, is more than 7 miles long. But Shallalah Saghirah may be the only community where a *qanat* is the sole water source.

There are no *qanats* in the surrounding villages in the Khanasser Valley—or that's what people told me. But the Bedouin had excavated underground cisterns up in the hills to provide water for their goats. Some still held water. The older men knew about them. I met Mohammed al-Issa, who took me to see some cisterns up a long gulley near the town of Habs. It was a stumbling climb over rocky terrain, past abandoned shacks built by shepherds who once brought their animals up here. But every few hundred yards there was a hole in the ground leading into a water-filled cistern.

At last we stood at the top of the gulley, beneath a single olive tree where, al-Issa warned, snakes rested in the shade. He told me the cisterns were all connected by a tunnel. And as we returned to the road, he pointed out a beautifully constructed stone channel that followed the slope down to a nearby village. It was dry now, but it must once have delivered water from the cisterns. The system was to all intents and purposes another *qanat*. So Shallalah Saghirah was not alone after all. But as we followed the channel, we noticed men loading pieces of its stone into a truck. They said they were taking it for building. Before our eyes, they were demolishing the *qanat*.

———

Qanats are one of the great engineering treasures of the Middle East, Central Asia, and North Africa. In these arid regions, rain falls only sporadically and mostly in the mountains, where there is little flat land or soil for farming. From there the water swiftly percolates underground, accumulating in the stony slopes that flank the eroding mountainsides. If the water surfaces at all, it does so in small fluctuating springs on valley floors.

It was the Persians, in modern Iran, who first learned to excavate these springs, chasing the water back into the hillside by digging horizontal tunnels. They found that the farther they dug, the more water the springs delivered and the more reliable the flow was. What probably began as a smart idea to get through a dry spell turned into the wellspring for farming societies across a huge swath of the Old World.

Most textbooks on water technologies make little mention of *qanats*, yet they are engineering marvels, dug straight and true, at a fixed gradient for their entire length, whatever the terrain above. And the scale of their construction is truly extraordinary. Syria is a minor outpost. In modern Iran, there are an estimated 50,000 *qanats*, mostly dug more than three thousand years ago during the heydays of the old Persian empires. The tunnels often extend for many miles into the hillside. Assembled end to end, they would reach two thirds of the way to the moon. If these structures were aboveground, they would surely be regarded among the great wonders of the world.

Once, *qanats* were the main source of water in Iran. Until the 1930s, dozens of them supplied the capital, Tehran. One gushed into the grounds of the British Embassy. Tabriz, Shiraz, and many other cities ran on *qanats*. One, supplying Isfahan, was more than 45 miles long. Some had such strong flows that the Persians built water mills along them to grind grain. As recently as the 1960s, their total discharge was around 485 million acre-feet a year, equivalent to the flow of twelve Nile Rivers. But modern water pumps have lowered water tables, drying up many *qanats*. The most recent estimate of productive use, made in 1998, was around 8 million acre-feet.

In many places they are now little regarded, and the skills that made them are dying fast. But in parts of Iran they still form the backbone of the water supply. The city of Irbil uses *qanats* that were dug by slaves of King Sennacherib, 2700 years ago. When an earthquake struck Bam in eastern Iran in late 2003, one of the first discoveries made by aid workers arriving in this remote town was that it was entirely dependent on *qanats* for water—and most of the tunnels had collapsed in the quake.

In the hot desert of central Iran, the ancient city of Yazd relies on *qanats* up to 40 miles long to bring water from the snow-covered Mount Sir and channel it beneath the city streets like a subway system, supplying water to underground public water houses, called *ab-anbars*. Yazd is a city of tiled mosques and carpet shops and ancient Persian wind towers, which ventilate houses through turrets oriented toward the prevailing valley winds. Many of the water houses have wind towers attached. They are oases of cool air, cool water, and well-being.

Wherever the Persians went, they took the secrets of *qanats*. The technology spread along the Silk Road to Afghanistan and China, where *qanats* are called *karez;* through Arabia, where they are *aflaj;* and along the north coast of Africa, where locals call them *fogarra*. *Qanats* watered Moorish Madrid and are still the mainstay of peasant agriculture in the Algerian Sahara, where the 50,000-acre Ouled Said oasis grows dates watered from *fogarra* dug into the nearby hills. Recent excavations have found hundreds of *qanats* beneath the Libyan desert, relics of the ancient civilization of Garamantes. They lie abandoned not far from the green pipes of Qaddafi's Great Manmade River. In Kurdish northern Iraq, the great city of Sulaimaniya still makes heavy use of them.

One spring morning a few years ago in Cyprus, I interviewed a man who claimed to be the last *qanat* digger on the island. To prove the point, sixty-five-year-old Yannis Mitsis took me to the orchards near his village, where he assembled a large wooden tripod, attached a pulley, and lowered himself down a hidden well into a tunnel that still fed water from the snow-clad mountains. The sun had already dried up the rivers in the lowlands, but his tunnel was running with water.

Mitsis described how as a young man he had helped excavate new tunnels, which are locally called *laoumia*. It was an ancient family occupation. And a dangerous one—his father had died years before when a rope broke. In the old days, he said, "The tunnels were the source of water here, and water was power." Nobody had dug a new tunnel since 1954, but he was kept busy doing repairs. Many orchards still relied on them. There was nobody to take his place, however. "My son is a criminal lawyer in London," he told me. "When I stop work, there will be nobody left."

In the hills of Palestine, there are tunnels very similar to *qanats* but shorter and excavated from hard rock. They are known locally as spring tunnels, and most villages on the West Bank have one somewhere. Many have been mapped by an Israeli geographer from Tel Aviv University, Zvi Ron. But most are used only as a last resort by poor Palestinians such as Ahmad Qot, whose story we heard in Chapter 18. "Most hydrologists don't know much about them," says Clemens Messerschmid, of the Palestinian Water Authority. "But I collect data from one just outside Ramallah. It delivers only 1.6 quarts a minute, but it is very reliable even in a dry summer, and the landowner still maintains it."

Almost everywhere, the *qanats* seem to be dying. And yet they have hydrological qualities that should be in great demand. They are self-regulating, because unlike electric pumps, they tap the aquifer only up to the limit of natural replenishment. As a result, if properly designed and maintained, they are, as they call them in Yemen, "unfailing springs." They keep going in all but the worst droughts. And they are a source not of conflict, like many pumped wells, but of social cooperation. The tunnels often emerge from underground at the home of the most important family in the village, but thereafter everyone is entitled to water. Typically, *qanats* have many owners and the water is divided into hundreds or even thousands of time-shares. In Ardistan in central Iran, the division of waters at one *qanat* goes back eight hundred years to when Hulaga Khan, the grandson of Genghis Khan, ordered the share-out.

Can *qanats* make a comeback? It is difficult to see a revival of traditional, manual construction. While their hydrology is all right, their sociology is all wrong. In Iran a whole caste of people low in the social pecking order had the task of excavation. It was dangerous and time-consuming work, and nobody has yet found a way of making it safe. But some water engineers believe that *qanat* digging could be mechanized to provide new, cheap, and permanent sources of water. Having seen ancient *qanats* in cliffs south of the Dead Sea, the Israeli hydrologist Arie Issar has proposed developing mechanized *qanat*-diggers there. Tunnels could tap ancient water beneath the Negev Desert that is uneconomical to capture in any other way, even with modern pumps. The idea has never been tested in practice, he says. Somebody should try it.

Meanwhile, there is a good case for making better use of existing *qanats* and for maintaining them and protecting them by placing restrictions on pumped boreholes nearby. Throughout the Middle East and beyond, there is growing interest in making the best use of these remarkable subterranean structures. In Oman, the government has paid for extensive repairs to six thousand *aflaj*, half the national stock. In the huge Turfan basin in the western Chinese province of Xinjiang, more than a thousand *qanats* dug during the Han and Qing dynasties were renovated in the 1990s. They provide a third of the district's water and take water up to 12 miles from beneath the snow-covered Bogdashan Mountains.

Aid agencies are also becoming interested. Often, abandoned *qanats* are the most viable new sources of water to supply refugees in zones of conflict. Mustang Valley, in the arid Pakistani province of Baluchistan, was once watered by 260 *karez*. After the valley was connected to the electricity grid and locals began to drill boreholes for water, the water table fell and the *karez* were abandoned. But during the drought in 2001, as Afghan refugees flooded across the border in the months after 9/11, teams of water engineers tapped the old *karez* to supply the refugee camps. In 2005, a U.S. government aid program in Baluchistan earmarked more money for *karez* rehabilitation.

When American soldiers went to Afghanistan in 2001, they developed a deep suspicion of tunnels in the hills. Many believe that Osama bin Laden escaped the U.S. dragnet in Tora Bora in the final days of the war by fleeing down *karez*, many of which are wide enough to accommodate small companies of men. Certainly in the 1980s, the tunnels made impenetrable hideouts for the U.S.-backed mujahideen during their guerrilla war with Soviet occupiers.

But American engineers were amazed to discover that *karez* were indispensable to Afghan farming. They may water as much as a sixth of the country's irrigated fields. If anything, their importance has grown in the past decade, as drought has dried up rivers and springs. In 2001, the aid organization Islamic Relief encouraged locals to renovate seventy-five *karez* in Helmand Province, where the Helmand River had run dry. And since the removal of the Taliban, Western aid agencies have decided to continue the work. The UN has proposed constructing small check dams in gullies to raise local water tables and bring *karez* back to life. In such inhospitable terrain, at least, these mysterious tunnels could be ready for a comeback.

X

When the rivers run dry...

we go with the flow

31

Learning to Love the Floods

Slavek the hippo enjoyed the worst floods in central Europe since the Middle Ages. He floated free from Prague Zoo and swam in the muddy waters that engulfed the heart of one of Europe's great cultural cities. He was happy, if a little hungry, by the time he was rescued as the waters of the Vitava River receded.

Nobody else much enjoyed the floods of autumn 2002. Central Europe is not the Mekong or the floodplains of Africa, where floods are good news, a necessary revitalization of vital ecosystems. In Prague, 40,000 people fled their homes, and only an enormous rescue effort saved the old town and its architectural treasures from total destruction. The cleanup bill was $2 billion. And there was worse damage downstream as the Vitava joined the Elbe, where the German city of Dresden also suffered badly.

So what had gone wrong? The rains were, of course, intense. But this was not a one-time event. It formed part of a recent pattern of surges of water out of central Europe's mountains. Reports of hundred-year floods somewhere in central Europe are now almost annual events. The Danube has suffered two in eleven years and three in fifty years. In March 2001, the rivers spilled over their banks in Hungary, Ukraine, and Romania, followed in July by similar inundations in southern Poland. The year before it was eastern Hungary and Serbia, and in July 1999 much of the Serbian capital, Belgrade, had to be evacuated as the Danube washed through it. In 1998 it was the Czech Republic, Poland, and Slovakia in July, followed by Slovenia and Romania in November. In six years, flood damage in mainland Europe cost the continent $30 billion.

The summer rains are getting fiercer, owing possibly to global warming. Glacial melting adds to the flow. But something also has gone badly wrong with the way Europe manages its rivers. Engineering intended to prevent floods is, it increasingly seems, conspiring to create them. We've seen how the operation of large dams can cause floods. But this problem extends to the entire management of a river system. All the major rivers that flooded in 2002 had been engineered specifically to banish floods. Their wetlands had been drained and their meanders straightened. The rivers were held in check behind high levees, and all impediments to having the water rush to the sea had been dynamited or chain-sawed from their path. But instead of eliminating floods, all this effort simply sped the water to the nearest bottleneck, where the floods became concentrated.

The strategy of rushing floodwaters to the sea, Europe's administrators were forced to conclude, has failed. "Flood peaks are higher and more damaging in places where wetlands and floodplains have been cut off from rivers, channeling more water into an unnaturally small space," Günther Lutschinger, a floodplain ecologist for WWF Austria, told me as authorities cleared up the mess in 2002. At the same time, many of the continent's mountains have lost most of their forests, reducing their capacity to absorb heavy rains.

In a single day at the height of the floods of 2002, most of central Europe received more than 7.8 gallons of rainfall for every square foot of land. Most of it surged within hours into the region's rivers, where 80 percent of the rivers' floodplains had been barricaded off by 600 miles of dykes, and where "improvements" had shortened the rivers' main channels by a quarter. The resulting floods came downstream in a great rush. And when the dykes failed, Prague and Dresden were engulfed.

The European Union is trying to improve forecasting of intense rainfall and to model how such rains affect river flows better. That may help cities prepare evacuation plans, but it won't stop the floods. Europe's crowded floodplains are at constant risk of inundation. So is it time for a new approach —time to learn how to love the floodwaters, to embrace them instead of trying to get rid of them?

There are two ways of beating floods. You can eliminate the water fast, draining it off the land and rushing it to the sea down high-banked rivers that have been reengineered as high-performance drains. Or you can encourage nature to hold on to it, letting it go only when the rains have stopped and the rivers are lower. Until recently, engineers preferred Plan A, the fast option. But however big they dug city drains, however wide and straight they made the rivers, and however high they built the riverbanks to keep the water on its path, the floods kept coming back to taunt them. From the Mississippi to the Danube, the flood-free future has failed to arrive. Dykes turn out to be only as good as their weakest link—and nature will unerringly seek that out.

By trying to turn the complex hydrology of rivers into the simple mechanics of a water pipe, engineers have often created danger where they promised safety and intensified the floods they intended to prevent. So now, more and more engineers are turning to Plan B. They are holding back the floods, capturing the water in fields and tearing down the banks and dykes and levees to give rivers back their floodplains. They are putting back the meanders and marshes to slow down the flow. They are even plugging up the drains on farms and in cities, encouraging the floodwaters to percolate underground. Rivers need room to flood, they say. And cities need to become more porous.

"The recent floods have provoked a completely new way of thinking," says the hydrologist Piet Nienhuis, of the University of Nijmegen in the Netherlands, a country that knows a thing or two about floods. "Rivers have to be allowed to take more space. They have to be turned from flood chutes into flood foilers."

Some call this soft engineering, going with the flow of nature. England's Environment Agency, which was given an extra $250 million a year to spend after floods in 2000 that cost $1.5 billion, is an advocate. It says, "The focus is now on working with the forces of nature. Towering concrete walls are out, and new wetlands are in." The soft engineers want to go back to the days when rivers took a more tortuous path to the sea and floodwaters lost impetus and volume while meandering across floodplains and idling through wetlands and inland deltas.

Take the case of the Rhine. Attempts to tame the river began in earnest in the nineteenth century, with "rectification" works undertaken by the German engineer Johann Tulla. Tulla's Rhineland rectification was the most spectacular of a series of great river-engineering projects across Europe. The very word "rectification" indicates the extent to which engineers' notions about recreating the world on their own terms had taken hold.

Until then, the upper reaches of the Rhine took a circuitous, wandering path across a wide floodplain of woods and water meadows. Between Basel in Switzerland and the German industrial city of Karlsruhe, the river split into innumerable branches that continually moved, disappeared, and reformed. The islands between the branches were covered by flooded forests and wet cattle pastures. During each spring flood, the silty Rhine water slipped into the forests, meadows, and fields.

It sounds idyllic, but the split channels and snaking route prevented the passage of all but the smallest boats, and building was difficult on shifting riverbanks. Even more disconcerting, the border with France, which followed the main channel of the river, moved whenever a flood passed. So Tulla forced the river into a single, well-defined, permanent channel. "As a rule," he declared, "no stream or river needs more than one bed." Nature never intended that this should be so, but Tulla's maxim has become the rule that almost every river engineer follows.

The rectification prepared the Rhine to become the great river highway of Germany. Along its banks grew industrial cities such as Mannheim, Koblenz, Cologne, and Düsseldorf. More than two thousand islands in southern Germany disappeared, and most of the sluggish backwaters and shallow gravel reaches disappeared. The new, straight upper Rhine was about 60 miles shorter than the old river, and it flowed a third faster.

But this was only a halfway house to the modern Rhine. Tulla's plan kept high dykes well back from the river, maintaining much of the floodplain. As time passed, however, farms, villages, and towns added their own dykes nearer the stabilized river, with the aim of protecting buildings and turning seasonally flooded pastures into arable fields. And during the twentieth century, an entirely engineered channel, complete with giant locks and a series of hydroelectric plants, finally cut off the upper Rhine from 60 square miles of its floodplain, leaving just a tenth of the original area available over which the river could flood.

The faster river sped barge traffic to the sea but also halved to just thirty hours the time it takes flood peaks to pass out of the Alps from Basel to Karlsruhe. After heavy rains, the peak flow on the Rhine now coincides with peak flows in tributaries such as the Neckar as they meet the main river. The resulting flood surge rushes downstream toward Bonn and Cologne, where peak flows are a third higher than before. Floods that previously were likely once every two hundred years can now to be expected every sixty years.

On top of that, the new, faster river scours the river's bed and banks much more fiercely. This has been exacerbated because the artificial embankments on the remade Rhine prevent the river from collecting silt from its former upstream floodplains, so it is much more able to erode material from the riverbed as it travels downstream. At Basel, the bed of Tulla's rectified river has fallen by 23 feet; at Duisburg it fell by 13 feet.

Tulla was known in his time as the "tamer of the wild Rhine," but his efforts helped turn the old, slow, silty stream that laid down alluvium as it went into a fast, silt-starved, and eroding river. Almost two centuries of efforts to tame the Rhine have made it wilder and more unruly.

———

Similar stories are told around the world—not least on the Mississippi, which drains the planet's second largest river catchment. The Spaniards who first arrived on the floodplain of the lower Mississippi in the sixteenth century thought better of building on it. They left great expanses of marshes and lakes to the Indians, who had long experience of coexisting with the floods. But the French settlers who followed felt no such constraints. They founded New Orleans right on the delta and called their marsh colony Louisiana, after their king.

New Orleans was always trouble. Water filled the basements of its buildings in the first months after its founding. Over the years, the French raised the natural levees ever higher in an effort to prevent floods. After an inundation in 1735, they built earthworks for 45 miles around the city. And after the United States bought the lower river as part of the Louisiana Purchase, slaves from dozens of cotton plantations, equipped with shovels and wheelbarrows, raised hundreds of miles of banks to speed "ol' man river" to the sea. But still it flooded.

In the mid-nineteenth century, Congress stepped in with the Swamp Land Acts, which handed ownership of unclaimed swamp to state governments.

They were allowed to sell the land to raise money for further levee building. This brought to the Mississippi a land-drainage fever that had already taken hold in Europe. Louisiana sold off a quarter of the state to farmers, who drained the swamps for plantations. But in so doing, they undermined the effect of the levee construction that the sales were funding, for in hydrological terms, the swamps were giant holding reservoirs for the river's natural floods. They spread the river's flow through the year, holding on to floodwaters during high flow and letting it seep back during low flow. Now barricaded off, they could do neither. The river floods grew higher, and the levees broke with increasing regularity.

Congress upped the stakes again in 1879, handing control of the river itself to the U.S. Army Corps of Engineers through the Mississippi River Commission. The corps's aim was to wage war on the river and finally "prevent destructive floods." But the river gave its reply by breaking the levees in 180 places in 1882. In 1927, despite ever greater expenditure by the corps, Mississippi floods submerged an area the size of Belgium, destroying 160,000 homes and every bridge for 1000 miles upstream of Cairo, a small Illinois town with a name as presumptuous as the engineers' science. That year, New Orleans was saved only when, recognizing the river's unstoppable force, corps engineers blew up a levee to draw off the flood.

Since 1927, some $7 billion more has been spent on raising levees. But separating the river ever more firmly from its floodplain has only increased the flows during major floods. The 1993 floods demonstrated that. Much of St. Louis was submerged beneath water that rose 50 feet above the normal level. The river rose over more than 600 miles of levees. Nearly five hundred counties featured in a presidential disaster declaration, and property damage was put at $12 billion. And then came Katrina. The hurricane that battered New Orleans and the Gulf Coast at the end of August 2005 came from the sea. But its most lasting impact came from storm surges into the Mississippi delta. Because the Louisiana wetlands have mostly been drained, the surge of water had nowhere to go. Water levels rose and breached New Orleans levees, engulfing a city of half a million people.

One of the river's nineteenth-century engineers, James Eads, uttered one of the great clarion cries of river engineering when he declared, "Every atom that moves onward in the river, from the moment it leaves its home among the crystal springs or mountain snows, throughout the 1500 leagues of its devious pathway, is controlled by laws as fixed and certain as those which direct the majestic march of the heavenly spheres...The engineer needs only to be insured that he does not ignore the existence of any of these laws, to feel positively certain of the results he aims at." In practice, the river has proved considerably more intractable. Mark Twain was perhaps closer to the truth when he observed after the creation of the Mississippi River Commission that "ten thousand River Commissions, with the mines of the world at their back, cannot tame the lawless stream, cannot say to it Go here or Go there and make it obey."

———

These brutal lessons are being widely learned. Plan B is now up and running as governments try to fend off the huge costs and political difficulties that can follow major floods. To help keep London dry, England's Environment Agency is breaking banks and reflooding 4 square miles of ancient Thames floodplain at Otmoor, outside Oxford. Farther downstream, its engineers have spent $180 million creating new wetlands to protect the boom town of Maidenhead, the ancient playing fields of Eton, and the deer parks of Windsor Castle.

Meanwhile, in one of Europe's largest river restorations to date, Austria is letting the Drava River back onto 40 miles of floodplain as it leaves the Alps. Its engineers are widening the riverbed and channeling the river back into abandoned meanders, oxbow lakes, and backwaters overhung with willows. The restored floodplain, they calculate, can now store up to 8000 acre-feet of floodwaters and slow up storm surges out of the Alps by more than an hour, thus protecting towns downstream all the way to Slovenia and Croatia.

The Dutch, for whom preventing floods is a matter of survival, have perhaps gone furthest. A nation built largely upon drained marshes and reclaimed seabed had the fright of its life in the early 1990s, when the Rhine twice almost overwhelmed it. In 1995, a quarter of a million people were evacuated. The resulting inquiries concluded that the rivers had been straightened and the land had been built on so much that there was nowhere for the water to go.

Rob Leuven, of the University of Nijmegen, says, "Government policy is now to allow rivers to take more physical space." Rivers are being allowed back onto their floodplains, and overspill areas of low-lying land called "calamity polders" have been set aside to lessen the impact of major floods. The Dutch are caricatured as always having their fingers in their dykes, but now they have given soft engineers the go-ahead to punch holes in the dykes. They are making plans to return up to a sixth of the country to soggy nature in order to protect the rest better.

Similarly, since 1995, the Germans have given top priority to finding ways of lowering the floods on the Rhine. Their target is to shave more than 23 inches off the flood peaks by 2020. Nobody knows if this can be achieved. But to start, they plan to reinstate some of the 500 square miles of floodplain lost on the lower Rhine. Drained and dyked fields will be replaced by reed beds and water meadows designed to flood every winter. Just twelve months after the great floods on the Elbe in 2002, the German environment minister, Jurgen Trittin, announced legislation that will "give our rivers more room again; otherwise they will take it themselves."

On a crowded continent like Europe, giving land back to the rivers is easier said than done, says John Handmer, of the Flood Hazard Research Unit at Middlesex University. A tenth of Europeans live or work on floodplains. The pressure is growing to ban all new developments on floodplains, but protecting those already there is best approached by finding ways to slow up water before it reaches the rivers. That should mean using subsidies from the European Union's Common Agricultural Policy to discourage farmers everywhere from "improving" their field drains. Poor drains that hold on to water longer are better for rivers. And, even more radically, it means redesigning the cities.

Modern cities could hardly be better constructed to create floods. They are concreted and paved and asphalted and culverted so that rains flow quickly into rivers. But the new breed of soft engineers wants our cities to become porous, so that they can capture and store that water instead. In Germany, Berlin is leading the way. The city's massive redevelopment since the fall of the Berlin Wall has been governed by tough new rules to prevent its drains from becoming overloaded after heavy rains. Harald Kraft, a city architect specializing in the new systems, says, "We now see rainwater as a resource to be kept rather than got rid of at great cost." Here is genuinely radical thinking. New ideas for flood protection begin to connect up to new ideas for harvesting the rain.

Take the giant Potsdamer Platz, a huge commercial redevelopment in the heart of Berlin. The city council has set a maximum limit for drainage from the site of just 1 percent of the potential runoff during a big storm. If the project doesn't meet the target, the drains will back up. Simple as that. So architects have designed the buildings to capture rainwater from the roofs. The water will flush toilets and irrigate roof gardens. Meanwhile, water falling to the ground will go to fill an artificial lake or percolate underground through porous paving. All told, the high-tech urban development can store a sixth of its annual rainfall and reuse most of the rest. It needs fewer drains. And it also needs fewer water-supply pipes.

New housing developments across Berlin are adopting similar technology. In the Zehlendorf suburb, rain from the roofs, gardens, and drives of 160 houses is collected to irrigate parkland. Not a drop of water leaves the area. Harzahn, a drain-free development of 1800 homes packed onto just 75 acres, features roads built of cobblestones to allow rainwater to percolate through the gaps into the soil beneath. Could this be done on a citywide scale? The test case could be Los Angeles. This is where the benefits of combining modern thinking on flood protection and water supply could really pay big dividends.

LA, as the song says, is a great big freeway. With hard, impervious surfaces covering 70 percent of one of the world's largest cities, drainage is a huge task. LA has spent billions of dollars digging drains and concreting riverbeds to take the water from occasional intense storms to the ocean. And still many communities regularly flood. The city engineers, locked in old thinking, want to spend $280 million raising the concrete walls on the Los Angeles River by another 6 feet. Meanwhile, with terrifying irony, the desert city is shipping in more water by pipe and canal from hundreds of miles away in northern California and the Colorado River in Arizona to fill its faucets and swimming pools and irrigate its golf courses. Is this sensible?

Like Phoenix, southern California has a self-image as a desert region that can revel in water. It has the biggest swimming pools, the highest fountains, and the most water-hungry agricultural crops. No problem. It prides itself on being able to pay. Marc Reisner titled his memorable book on the state's water politics *Cadillac Desert*, a phrase that sums up the mindset. But this strutting hides a serious misreading of the state's geography, which sounds like bad planning. "In LA we receive half the water we need in rainfall, and we throw it away. Then we spend hundreds of millions to import water," says Andy Lipkis, an LA environmentalist. "We should be catching our own rain before trying to buy other people's."

Lipkis, along with citizens' groups like Friends of the Los Angeles River and Unpaved LA, want to beat the urban flood hazard and fill the faucets at the same time by keeping the floodwater in the city. They call their dream the "porous city." They are working with schools and other places to convert asphalt areas into soft ground to soak up rain. And they believe it is a realistic proposition to do this across the city.

It could happen. The city authorities have recently established a watershed management division, and in 2004 they launched a $100 million project to road-test the porous city in one poor flood-hit community in Sun Valley. The plan is to catch the rains that fall on thousands of driveways and parking lots and rooftops in the valley before it gets to the drains. Trees will soak up water from parking lots. Homes and public buildings will capture roof water for irrigating gardens and parks. And road drains will pour water into old gravel pits and other leaky places that should recharge the city's underground water reserves. It's a direct replacement for a planned storm drain. The result, the city hopes, will be less flooding and more water for its inhabitants.

Plan B holds that every city should be porous and every river should have room to flood naturally. It sounds expensive and utopian, until you realize how much we already spend trying to drain our cities and protect our floodplains—and how bad we are at doing it. Katrina only underlined that. Plan B, going with the flow of the river, really is the better way. The question now is whether that can form the basis for a wider philosophy for managing rivers and the water cycle.

32

Freeing Saddam's Captives

In the aftermath of the first Gulf War in 1992, Saddam Hussein began draining the bulrush marshes of southern Iraq, home of the Madan, or Marsh Arabs, and sometimes described as the original Garden of Eden. Now, in the aftermath of the second Gulf War, wetland scientists are joining with Madan and Iraqi exiles in the United States to rehabilitate the marshes. They are planning the largest and most ambitious recovery of a lost wetland ever attempted. The first phase of the rehabilitation was begun by returning Madan in the months after Saddam was toppled; it cost little more than a few hours with a bulldozer. To complete the job might cost tens of millions of dollars. But if it can be achieved, it could be a model for reviving many other natural water reservoirs as the world fights to stave off water shortages and to remake drying rivers.

When watching TV images of U.S. forces crossing the desert of southern Iraq during the spring of 2003, few realized that until a decade before, most of the land between Basra and the Euphrates bridges at Nasiriyah was marsh. Back then, the tank lanes were covered in lakes and a network of waterways hemmed in by banks of reed beds. In this strange enclosed world, the great rivers Tigris and Euphrates nourished a flourishing ecosystem twice the size of the Everglades. It sustained huge numbers of fish, large populations of birds, and rare animals such as the smooth-coated otter.

But this was not a virgin wetland. It was a highly productive natural factory that had given the Madan everything they wanted for five thousand years.

Tigris and Euphrates Rivers

Upwards of half a million people grew rice, caught fish, hunted otters and birds, gathered reeds, and grazed their water buffalo in a world that had not changed much since it was immortalized in romantic simplicity by Wilfred Thesiger's 1960s book *The Marsh Arabs.*

But Saddam changed all that. In 1993 and 1994, he dug drains to empty the marshes of water, diverted the two great rivers away with huge canals, and raised embankments to prevent future flooding. By the end of the decade, more than 90 percent of the marshes were gone. Their loss was "a major ecological disaster, comparable to the deforestation of Amazonia," said Klaus Toepfer, the director of the UN Environment Program. It was also a crime against the Madan and a major loss to people in Iraq who had never been near the marshes. Before the engineers moved in, the marsh waters had provided 60 percent of Iraq's fish and much of its rice.

Saddam always insisted that his vast engineering projects on the marshes were intended to boost farming. They would, he said, reclaim the land for farming and increase Iraq's crop production by up to a half. It was not a self-evidently preposterous claim. This is the same reason that most engineers and their political and financial backers have always given for draining marshes. And a similar fate for the Mesopotamian marshes had been proposed by British engineers fifty years before.

On a shelf in the Institution of Civil Engineers in London, I found a report called "Control of the Rivers of Iraq." It was published in 1951 by the Iraqi Irrigation Development Commission and written by a senior British colonial engineer of the day, Frank Haigh. In one chapter, Haigh proposed diverting the Euphrates into a "Third River" that would reach all the way from Baghdad to the Persian Gulf, bypassing the marshes. And he also proposed installing sluices to restrict the flow of the Tigris into the marshes. Haigh's plan was to capture the water then "wasted" in the marshes and use it for irrigation. As I read the report, back in 1993, as Saddam's engineers were getting down to the job, I realized that Saddam's scheme was essentially a carbon copy.

In fact, construction of part of the Third River had begun under British supervision in the 1960s. "We did the early design work for the Third River," Bill Pemberton, at the British firm of consulting engineers Mott MacDonald, later told me. "We built our bit, about fifteen miles. The rest they now seem to have done themselves." But whatever Saddam's long-term aims for irrigation and whatever blueprints he drew on, most agree that his immediate political purpose in draining the marshes was to empty them of the Madan, who had been prominent in the rebellion against him in the months after the first Gulf War.

In that he succeeded. A marsh population of half a million was reduced to fewer than 80,000. Madan refugees were scattered across the Middle East, and their homeland was turned into a salt-encrusted desert dotted with burned-out villages and poisoned with the pesticides that Saddam's soldiers poured into the marshes to kill any remaining life.

Before Saddam, the marshes were by some way the largest area of wetland in the Middle East. They comprised three distinct areas—the Hawr al-Hammar marshes south of the Euphrates; a large complex of permanent lakes and marshland between the Euphrates and the Tigris known as the Amara marshes; and the al-Hawizeh marshes, which stretched east from the Tigris into Iran. The last of these survived Saddam's engineering the best because it was partly fed by the Karkeh River flowing out of Iran. But by the time of the second Gulf War, that too was suffering badly as Iran filled a giant new dam on the river as part of a big agricultural development project.

The Karkeh rarely crosses the border into the marshes anymore. According to UNEP's Hassan Partow, lakes that were permanent features as recently as 2000 have disappeared and the al-Hawizeh marshes have become "a landscape of parched beds, withering reeds, and sterile salt flats." The remaining wet areas are likely to disappear altogether before the end of the decade.

———

As the second Gulf War drew to a close, an unexpected defender of the Mesopotamian marshes emerged. Azzam Alwash is an Iraqi exile living in California, whose father had been an irrigation engineer helping outsiders grow rice on the edge of the marshes. On the face of it, you might expect Alwash to favor draining more of the marshes. But, as he told me in an interview as U.S. forces charged across his old homeland in the direction of Baghdad, he was a determined defender of the marshes and wanted to form a scientific SWAT team to bring back the water.

As a boy, he explained, he had gone duck-shooting with his father in the marshes on weekends. He wanted to rehabilitate them so he could one day go back. He and his American wife, Suzie, a wetland ecologist, gathered scientists from around the world. They called their effort the Eden Again Project, and it had the support of the U.S. State Department, UNEP, the World Bank's Global Environment Facility, and eventually the interim Iraqi government.

But the scientists had a warning for anyone trying to bring back the water in a hurry. The desiccated soils on the former marshes were fragile, they said. They were caked with salt left behind as the marshes dried up in the sun in the mid-1990s. The danger was that as the land was reflooded, the accumulated salt would flush through former freshwater marshes, poisoning them. The freshwater marshes were the most important ecological feature of the whole, they said, being home to most of its rarest species. "The one thing we cannot do is just plug up Saddam's canals and let the water flow back into the marshes," said Tom Crisman, the director of the University of Florida's Center for Wetlands and a veteran of rehabilitation projects in the Everglades.

In fact, that was precisely what happened in the months after Saddam fell. Some of the Madam moved back and, with help from local engineers from the reconstituted Ministry of Water Resources, began tearing down embankments, breaking dams, and opening sluices to allow water back onto the marshes. At the time of writing, it is not entirely clear whether the scientists' forebodings have proved correct. At the height of the spring floods in 2004, a quarter of the marshes had reflooded. In places, reeds had grown 10 feet high, fish were living among them, migrating birds had returned to nest, and some Madam were using them to rebuild their traditional houses. But in other places there was little sign of recovery, perhaps because of salt problems.

The interim Iraqi government said it wanted to restore as much of the marshes as it could. At the Stockholm Water Symposium in late 2004, the Iraqi water minister, Jamal Rashid, told me, "Maybe we won't be able to restore 100 percent of the marshes, but 70 or 80 percent would make me quite pleased." That is a big task. "The job can be done, but it will have to be a phased approach, starting with areas near the main rivers and then working into the center of the marshes," said Crisman. And it could cost half a billion dollars.

Will it happen? That too is unclear. For one reason, after the first burst of environmental enthusiasm, other ministries with different agendas have been taking an interest in the marshes. There are competing demands, not least the possibility that large amounts of oil lie beneath the reemerging reed beds. But the big problem now that Saddam and his engineers have departed is water. Will there be enough to refill the marshes?

Like wetlands the world over, the Mesopotamian marshes are threatened by increasing demands for water upstream. Less and less water has been flowing down the Tigris and Euphrates in the years since Saddam sabotaged the marshes. Indeed, it is a moot point to what extent Saddam killed the marshes and to what extent new dams upstream did the dirty deed. Most of those dams are in Turkey, the source of most of the rivers' flow. Dams like the giant Ataturk Dam on the Euphrates have cut flows on the two rivers into Iraq by a fifth. And, perhaps even more damaging for the marshes, they have eliminated the peak spring floods, which inundated the widest areas of marshes. As on other great rivers, from the Mekong to the Logone, the flood-season flows sustained the greatest biological productivity, to the benefit of both wildlife and the humans who tap the natural resources.

"With more dams being built all the time, the whole region is in a water crisis," said Hassan Partow. Since the end of the war, UNEP has been trying to broker a deal among the nations on the rivers that feed the marshes to share water among themselves and to engineer a spring flood pulse into the marshes. But the signs are not good. Turkey has been especially hostile. And Iran has refused to budge on its ambitions to take most of the water out of the Karkeh River before it reaches the al-Hawizeh marshes.

In the short term, if Iraq wants to recreate the marshes, it may have to go ahead with what water it has. "Even with the present amount, it should be possible to do a lot," Partow told me. "The key is to reestablish some flows of freshwater through the main marsh areas. After that there may be a choice between creating one large shallow saline wetland and sealing off some salty areas to create smaller areas of fresher water." But a full recovery will require reconsideration of the management of the rivers by all four countries.

Meanwhile, Alwash is waiting and dreaming. "I used to go hunting and picnicking with my father on the marshes every week," he told me. "You could travel by kayak for hundreds of miles from my home in Nasiriyah. My dream is that one day soon I will be able to go again with my own children." If it can be done, it could be of huge importance for remaking the world's rivers.

33

More Crop Per Drop

Can Pepsee save the world? Pepsees are small tubes made of light, disposable plastic and designed to encase individual ice candies—lollipops, or whatever you choose to call them. They are manufactured all over the world and are just about the most disposable product imaginable. In India they are made under the Pepsee brand name. Millions of roadside vendors across the land buy the polyethylene tubes, which come in long rolls that can be torn off at perforations stamped into the plastic every 8 inches or so.

Sometime around 1998, somewhere in the Maikal hills of central India, someone—perhaps a farmer with a sideline of selling ices—started using Pepsee rolls for another purpose: to irrigate the fields. The farmer had discovered that the rolls of plastic tubing made perfect cheap conduits for distributing water to plants. Shilp Verma, of the International Water Management Institute in India, says, "It is not very clear how and exactly where the innovation first started, but it spread among farmers like wildfire." The farmers took the rolls, laid them down rows of crop plants as close as possible to the roots, and poured water in at one end. As the water ran down the tubes, it dripped through the perforations to water the plants.

At the start, says Verma, there was one problem. Algae grew in the wet, sunny environment of the tubes. But Indian innovation was equal to the task. Wise to the new use for their product, the manufacturers produced an even cheaper version made of black recycled plastic and known to farmers as the Black Pepsee. With light excluded, the algae problem was solved. This startlingly successful exercise in lateral thinking has produced the first dirt-cheap method of providing drip irrigation for poor farmers.

Farmers across the world have traditionally irrigated their fields by pouring on water indiscriminately. Some reaches crop roots; most does not. The result has often been waterlogged soils, a buildup of salt, and growing water shortages. At a conservative estimate, two thirds of the water sent down irrigation canals never reaches the plants it is intended for. Technologists early in the twentieth century got to work on the problem. First they came up with the sprinkler, which doused fields in a fine spray of water from a central pivot. That saved the worst excesses of overirrigation but was equally indiscriminate and lost a lot of water to evaporation. It was also expensive and required energy to spray the water around.

The next advance was drip irrigation. Various people claim this idea for themselves, but credit is normally given to an Israeli engineer called Symcha Blass, who retired to the Negev Desert in the early 1960s and, as he told it, one day noticed how a large tree grew in the desert because it was right next to a slowly dripping faucet. He mused on the matter and subsequently invented a narrow tube that could deliver water under pressure and drip it close to the roots of plants. He filed a patent in Tel Aviv in 1969. The trick, as much as anything, was timing. His idea coincided with the development of plastics that made such a system economical for the first time, and with the growing realization in arid countries like Israel that water was in increasingly short supply.

Drip irrigation can take many forms. Mostly it is high-tech, with water pumped down pipes under pressure and sent into side pipes from which sophisticated "drippers" deliver it to roots. Such systems can include flow meters, pressure gauges, and even soil-moisture gauges to optimize delivery and keep losses to a minimum. Today large farms in California, Tunisia, Israel, and Jordan specialize in such systems. In Jordan, drip irrigation has reduced water use on farms by a third while raising yields. Israeli farmers have raised water productivity fivefold in the past thirty years through a mixture of drip irrigation and the recycling of treated urban wastewater onto their fields.

The technology should have taken off. But it hasn't. It remains virtually ignored by the mass of small farmers in poor countries, who face some of the worst water shortages. In India, for instance, despite strong government promotion and subsidies, drips irrigate fewer than 1 percent of fields. "It is still largely seen as a technology for gentleman farmers," says Verma. This is perhaps not surprising, since the full kit can cost more than $800 for one acre, and even stripped-down systems developed for poor countries cost at least $200.

Another reason is that most farmers in most places at most times get their water at heavily subsidized prices—a tenth of the real cost is typical everywhere from India to Mexico and Pakistan to California. And when farmers pump water from beneath their fields, they pay no more than the cost of a pump and electricity, which itself is usually highly subsidized. Except in an immediate crisis, there is little incentive to save water. So what will turn the tide? More realistic pricing of water would certainly help. But so too would appropriate technology. And that means going back to basics.

Drip irrigation, it turns out, is not a brand-new technology. At least two thousand years ago, Chinese farmers were making small holes in earthenware pitchers and burying them in the soil. They would go around their fields every few hours to refill the pitchers, which then simply leaked water into the root zone. This indigenous system is also known in India and parts of Africa and the Middle East.

Indian farmers also have a tradition of irrigating from hollow bamboo tubes pierced with holes and laid out along fields. More recently they have adapted the bicycle inner tube, making a few extra punctures to allow a steady drip of water into the ground. But you can get your hands on only so many inner tubes, so now the Pepsee is spreading fast across India. No wonder it took off. A 2-pound roll of the stuff can be bought for fifty rupees (about one dollar). Even allowing for the necessary plumbing to deliver water, the overall cost —around 1000 rupees, or $23, an acre—is less than a tenth of that of even the cheapest conventional systems.

In the face of escalating water problems, agricultural researchers all over the world are adopting a new philosophy of "more crop per drop." It is a huge turnaround from the green-revolution days, when the philosophy might reasonably have been called "the more water, the better." Raj Gupta, the director of a large agricultural research center run by the International Maize and Wheat Improvement Center in the heart of Delhi, says, "For the first time, we are starting to measure crop yields in terms of the tonnage produced for a given amount of water, rather than a given amount of land."

Gupta holds that cheap drip-irrigation systems like the Pepsee could save

India from running out of water. To go with it, he has a whole group of changes to current farming methods. They include planting crops on raised beds, reducing plowing, and leveling the land to prevent waterlogging. Together, he believes these approaches could reduce water use across northern India, the breadbasket of the country, by a third or more, and save millions of farmers in the south from penury as the groundwater bubble bursts. And what will work in India should work in China and Colombia, Mexico and Mali, Libya and Lesotho.

Rice is an early target for the "more crop per drop" crusade. The world's most popular grain consumes more water than any other food crop, typically twice as much as wheat for every ton of output. More than a third of all the water abstracted from rivers and groundwaters on the planet goes to irrigate rice paddies in Asia. But with half of the world's rice grown in water-stressed China and India, that level of production cannot continue.

Experiments in India and at the International Rice Research Institute in the Philippines show that rice can be grown with much less water. The trick is to abandon the traditional method of growing seedlings in nurseries and then transplanting them into paddies that have to be kept flooded. Much less water is needed if they are instead planted directly into muddy soil. Some varieties can already be grown this way. If those varieties were more widely adopted, they would cut water use by a fifth—and require less labor into the bargain. Other crops offer similar benefits from water-saving methods of cultivation.

Researchers say this new approach amounts to a "blue revolution" to follow the green revolution. The Washington-based International Food Policy Research Institute (IFPRI), which is part of a network that also includes the international water management and rice research institutes and Gupta's maize and wheat center, says it is vital that technologies like drip irrigation be more widely adopted. Together, they could bring cost-effective global savings in water use of more than a fifth.

Already governments in water-short regions are eager to take up the challenge. Centrally planned China is in the forefront. Unable to wait for Yangtze water from the south-to-north project, Shandong Province on the North China plain—where the people get whatever water is left in the Yellow River after every other province has had its fill—responded to water riots in 2000 by agreeing to spend $6 billion on water conservation on farms.

Meanwhile, there is plenty to be done at plant-breeding stations. Most modern high-yielding plant varieties are hopeless water guzzlers. Only now are breeders turning their attention to producing varieties that use water efficiently. New rice varieties, the researchers say, could one day cut water use by as much as half. But another step is required. There has to be a reconsideration of where crops are grown. We need to fit the crops grown to the availability of water rather than attempting to do things the other way around. Many parts of the arid world simply need to give up growing water-guzzling crops.

Cotton is perhaps the worst offender. The fiascos in Central Asia are the most obvious case in point, but most of the cotton-growing in the world is destroying rivers. The cotton growers of the Indus should be halted. Many researchers go further. Gupta says that much of northern India should stop growing rice altogether and switch to wheat and corn. Similarly, the dairies of Gujarat should be weaned off cattle fed on irrigated alfalfa in the state. In China, many argue that the country should shift its entire farming effort from the arid north, where the Yellow River is running out, to the wetter south. Better that, they say, than spending tens of billions of dollars to take the water north.

Mark Rosegrant, at IFPRI, says the effects of all this could be profound. Cutting abstractions for agricultural irrigation water could leave an estimated 800 million acre-feet of water in the world's rivers each year. "Many planned dams will be cancelled," he says. From the Rio Grande to the Yellow River, the Indus, and the Nile, dried-up rivers could resume their flows. A start could be made on reviving the Aral Sea.

———

For most countries of the world, agriculture is the biggest user of water and the biggest cause of water shortages. And that is where the big solutions lie. But we can all contribute. Take the toilet. In most homes it is responsible for a third or more of all the water use. This does not need to be so. In the United States, the amount of water used to flush the nation's toilets has been cut by three quarters in the past two decades, thanks to bigger valves and S-bends that allow a shorter, faster, more efficient flush. The standard flush is now 1.6 gallons rather than 3.4 gallons. (This insured a substantial decline in indoor water use in American homes since around 1980, which, sadly, has been more than wiped out by increased outdoor use, especially for sprinkling lawns.) Similar savings can be made by redesigning everything from shower units and faucets to public urinals and industrial processes.

Unfortunately, not all countries take this approach. One British regional water company found that modifying every toilet in London would save more water, and be cheaper, than building a new reservoir on the Thames. But the company wants the reservoir anyway.

Water utilities that encourage their customers to save water would do well to put their own houses in order. In most of the world's cities—from London to Nairobi to Shanghai—between a third and a half of all the water put into the mains disappears through leaks before reaching its customers. In Africa, people in the countryside walk for miles to collect water while city water pipes leak half their contents into the ground. Yet cities with active programs to find and repair leaks have found that saving water is cost-effective. Singapore, which has to buy most of its water from neighboring Malaysia, has taken a tough approach and got its leaks down to 5 percent—probably a world record.

As we saw in chapter 26, we have to be careful how we treat water savings. Somebody downstream may rely on leaking mains and overirrigated fields for their water supply. But even so, if we take the trouble to use the water-saving technologies available, we can save water in the way we are learning to save energy and recycle waste—for the good of our planet as well as our pockets.

34

Water Ethics

The task of providing water for the planet's growing population is huge. Even basic needs for the existing population still go unmet. More than a billion people in shantytowns and remote villages across the poor world have no access to reliably clean drinking water. Most are in sub-Saharan Africa and South Asia. But we should not forget that there are hydrologically dispossessed people even in the rich world; they include the inhabitants of *colonias* on the U.S. side of the Mexican border.

The World Summit on Sustainable Development in Johannesburg in 2002 promised to provide for at least half of the dispossessed billion by 2015, by insuring their basic human right to clean water. Fulfilling that promise will require making new water connections for 125,000 people every day. Even if the pipes can be laid and the pumps installed, nobody is sure whether there will be enough water to do the job—especially if the newly connected then all install toilets that require flushing.

Tony Allan, of the School of Oriental and African Studies in London, says, "The key to avoiding catastrophic water shortages is bringing people out of poverty." He has a case. By and large, the people who are short of water are the poor. Even in desert countries, those with money can get water. Sufficient money can almost always deliver sufficient water for immediate needs—even if it requires a desalination plant or a ludicrously long pipe.

Already, across rich and poor countries alike, water is flowing uphill to urban users, who can almost always outbid farmers and the rural poor. Water demand for domestic and industrial use is expected to increase by two thirds by 2025. And in most places, one way or another, that demand will be met. But money thrown at problems often produces the wrong solution. The World Bank and water engineers said in Johannesburg that it would cost $180 billion to get clean drinking water and basic sanitation to those who don't have it, or about $100 per person. They meant that it would cost that much using conventional Western solutions, like large dams, water-treatment plants, pipe networks, and sewer systems.

That kind of money is unlikely ever to be found, any more than the world's poor farmers will be provided with high-tech drip-irrigation systems. But grassroots aid groups insist that the job can be done at a tenth of the cost with simpler technologies, like wells and rainwater harvesting and simple latrines. That must be the strategy.

If basic human needs for water can be met in most places at most times, what about providing enough water to grow the crops to feed ourselves? That is a far harder nut to crack. The amounts of water involved are much greater. Farmers today get two thirds of the water taken from the world's rivers and underground reserves, but that proportion is sure to decline as urban and industrial customers—including those for a handful of cash crops, such as cotton—take precedence in the marketplace. Food farming will always be the poor relation when water is in short supply.

By 2025, economists say, water scarcity will be cutting global food production by 385 million tons a year. That is more than the current U.S. grain harvest and the equivalent of a loaf of bread every week for every person on the planet. For hundreds of millions of people, that disappearing loaf may be the only one they get.

As the rivers run dry, more and more irrigation canals run dry too. And underground reserves cannot make up the difference. We are already living on borrowed time by mining the aquifers—dipping into the slow, often largely unrenewable water cycle to top up the fast, renewable cycle. We will live to regret this, and if we don't, our children will. Up to a billion people are today eating food grown using underground water that is not being replaced.

Another safety valve is importing "virtual water" in the form of food. Many nations already do this to replace crops they would once have grown for themselves. In the next two decades, water scarcity will shift where the world's food is grown, away from water-poor countries in North Africa, the Middle East, and much of Asia. Provided these countries have the money to buy virtual water, they may get by for a while. But there are limits to this: we cannot all import virtual water, even if we have the money. The real water has to be somewhere for the crops to be grown.

Perhaps the most telling case today is China. Historically, the world's most populous nation has almost always fed itself. But today, increasing water shortages are pushing China to import food in a big way for the first time. And such is the country's size that this is already affecting world food security. Strategic grain reserves are emptying and world grain prices are rising. Meanwhile, the world's second most populous nation, India, is filling its granaries by plundering diminishing underground water reserves. That cannot go on.

The agronomist Lester Brown asked in the title of a book on Chinese agriculture a few years ago, "Who will feed China?" But increasingly, as the rivers run dry, we have to ask, Who will feed the world?

———

The good news is that we never destroy water. We may pollute it, irrigate crops with it, and flush it down our toilets. We may even encourage it to evaporate by leaving it around in large reservoirs in the hot sun. But somewhere, sometime, it will return, purged and fresh, in rain clouds over India or Africa or the rolling hills of Europe. Each day more than 800 million acre-feet of water rains onto the earth. Water is the ultimate renewable resource. And there is, even today, enough to go around. The difficulty is in insuring that water is always where we need it, when we need it—for all 6.5 billion of us.

As the late American hydrologist Robert Ambroggi put it three decades ago, "The problem facing mankind is not a lack of fresh water, but a lack of efficient regimes for using the water that is available." The fact that we are so bad at managing water at least shows that the potential for doing better is high. The solution in most cases is not more and bigger engineering schemes. It is not south-to-north projects or river-interlinking projects or giant desert canals or megadams. Such projects are hugely expensive, and many are the cause of as many problems as they solve. They are, I believe, at the heart of our current inefficiency.

To manage the water cycle better, we have to give up the idea that water has to be extracted from nature and put inside metal or behind concrete before it can be used. We have to treat nature as the ultimate provider of water rather than its wasteful withholder. We must learn to "ride the water cycle" rather than replace it.

We have to treat water as a precious resource rather than something that just falls from the sky. We have to find the "more efficient regimes" of using the water cycle that Ambroggi spoke of. That certainly means doing better science and investing in a "blue revolution" to bring the old green-revolution crops in line with hydrological realities. But beyond that, we need a new ethos for water—an ethos based not on technical fixes but on managing the water cycle for maximum social benefit rather than narrow self-interest.

Access to water, most of the world agrees, is a human right. So water cannot always flow toward money, whether up hill or down dale. These new priorities will often mean going back to ancient ways, such as harvesting the rain where it falls. There is huge potential for this, and it will very often make much more sense than continuing with the twentieth-century obsession with large dams and mass transport of water in pipes and canals. My one technological bet for the twenty-first century is that rainwater harvesting will resume its preindustrial place in water management in many countries, providing local water to meet local needs.

But the new ethos will also harness modern methods and ideas. It will adopt high-tech irrigation to provide "more crop per drop" while recognizing the realities of finite water resources inside closed river basins. And it will explore new ways to combine different parts of the water cycle. Why build huge dams when you have natural reservoirs right beneath your feet?

One image of hope is the disciples of Dada and others in Gujarat who are pouring monsoon water down their wells. Why not develop this concept on a much larger scale? Proposals for the large-scale diversion of the monsoon floodwaters of the Ganges into the aquifer beneath its plain could make sense. The best way might be to pour river water into unlined irrigation canals and let it seep underground, thus turning conventional notions of irrigation efficiency on their head.

Also underground, we might revert to modern versions of *qanats* as a means of maintaining water tables and reducing pumping. The *qanat* is a classic technology for the communal management of water. It has fallen by the wayside not through any hydrological failing but because it does not easily fit with the contemporary vogue for private ownership of resources. We need to relearn some of the old lessons of sharing if we are to manage water better.

We must realize too that water has to be given back to nature. The environmentalists' case for insuring "conservation flows" in rivers and on wetlands is unanswerable. This is not an optional extra for the "tree-huggers." Nature's free services in maintaining fisheries, protecting against floods and drought, cleaning pollution, delivering free irrigation on floodplains, watering valuable tourist sites, and much else are just too valuable to be lost.

The new thinking means going with the grain of nature. To insure our water supplies and protect ourselves against damaging floods, especially in an uncertain greenhouse world, we will often have to tear down dams and dykes, recharge the underground reservoirs, and remake the rivers. We will certainly have to pour less concrete—a twentieth-century solution to a twenty-first-century problem.

The twentieth-century view that the world can feed itself only by artificial irrigation of huge areas of the developing world will also have to go. It is hubris we cannot afford. This kind of irrigation has always been a fantastically expensive and inefficient way of growing crops. It should be a last resort. That is why bodies like the World Bank have virtually stopped funding it. We seem to have forgotten that direct rainfall onto soil is, in Allan's words, "what makes possible well over half the world's crop production." Rain will continue to water most of our crops, and we can recognize that by adopting a global strategy for encouraging rainwater harvesting.

Luckily, there are already places where all this is starting, fitfully, to happen. On the plains of India, farmers are taking inspiration from swamis and scientists to capture the monsoon rains on their fields and in their aquifers. In Machakos, similar techniques have turned back the advancing desert while filling stomachs and swelling granaries. In Los Angeles, people talk of turning the most paved urban area on the planet into a "porous city" that can catch the rain, banish floods, and become self-sufficient in water. Similar smaller-scale schemes have begun in European cities. In Iraq, the inhabitants of the Mesopotamian marshes are returning to tear down Saddam's dykes and plug his drainage canals, remaking the wetlands that once gave birth to the story of the Garden of Eden.

———

Rivers are sacred in most religions. For Buddhists, the gods live at the center of the universe, where the great rivers of their world—the Ganges, the Indus, and the Brahmaputra—rise. Hindus go on pilgrimage to drink the holy waters of the Ganges, and—further downstream, thankfully—they cast the bodies of their dead into the same river. Christianity depicts humanity beginning in the Garden of Eden, where fountains fed rivers that watered the world. And in recognition, Christians baptize the faithful in water. We sing of the River Jordan and the waters of Babylon. The sources of water are everywhere revered. Australian Aborigines hold the waterholes and billabongs of the outback to be sacred places, physical manifestations of the process of creation. The Japanese mark the start of rivers with Shinto temples. Africans pray at springs in the depths of their sacred groves. Even Europe is peppered with ancient holy wells.

Rivers are symbols of nationhood, too. Like nations themselves, they are always present but always changing. Ol' man river, he just keeps rolling along. The Yellow River is China's joy and sorrow. Moses parted the waters of the Nile to save his people. The Canadians will trade almost anything except their most abundant resource: water. Russian nationalists in the 1980s rose in anger at the prospect of delivering Siberian water to their fellow socialist republics in Central Asia. The Thames in England and the Murray in Australia, the Loire in France and the Three Gorges in China, trout streams and salmon rivers and even crocodile swamps—all tug at the heartstrings of nations.

But we need new versions of these totems and myths, versions that recognize that the rivers may not be so permanent after all—that without conservation and management, they could run dry. Ol' man river, he just might stop rolling along. It is hard to say how or where these totems might emerge. As hard, perhaps, as predicting the emergence of the Pepsee as a vital irrigation technology. Some things just have to be allowed to happen. But for a start, Spanish opponents of plans to divert the Ebro River were on to something when they called for a "new water ethic" that cherishes water and respects the river.

That is something that Chinese rain harvesters and Indian water priests will understand. So will fishermen and environmentalists and the campaigners for "porous cities." It is an ethic that recognizes that rivers are more than just sources of water—more than the feedstock for irrigation canals and hydroelectric power stations. It recognizes that rivers provide fish and silt and recharge for underground reserves; that water purges and purifies; that there is virtue in flood pulses and in the mixing of land and water on a river's floodplain.

And it requires us to find ways of storing water without wrecking the environment, of restoring water to rivers and refilling lakes and wetlands without leaving people thirsty, and of sharing waters rather than fighting over them. It requires us to go with the flow. And to do it before the rivers finally run dry.

Index

Karen Nov. 19